Rethinking Evolutionary Psychology

Rethinking Evolutionary Psychology

Andrew Goldfinch
London School of Economics, UK

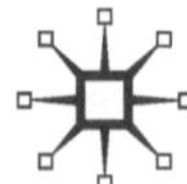

Softcover reprint of the hardcover 1st edition 2015 978-1-137-44290-1

First published 2015 by
PALGRAVE MACMILLAN

Palgrave Macmillan in the UK is an imprint of Macmillan Publishers Limited, registered in England, company number 785998, of Houndmills, Basingstoke, Hampshire RG21 6XS.

Palgrave Macmillan in the US is a division of St Martin's Press LLC, 175 Fifth Avenue, New York, NY 10010.

Palgrave is the global academic imprint of the above companies and has companies and representatives throughout the world.

Palgrave® and Macmillan® are registered trademarks in the United States, the United Kingdom, Europe and other countries.
ISBN 978-1-349-49522-1 ISBN 978-1-137-44291-8 (eBook)
DOI 10.1057/9781137442918

This book is printed on paper suitable for recycling and made from fully managed and sustained forest sources. Logging, pulping and manufacturing processes are expected to conform to the environmental regulations of the country of origin.

A catalogue record for this book is available from the British Library.

A catalog record for this book is available from the Library of Congress.

Typeset by MPS Limited, Chennai, India.

Contents

Acknowledgements

Roman Frigg is owed a great debt of gratitude. As Director of the Centre for Philosophy of Natural and Social Science, Roman was the first to suggest that I write a book on my conceptualisation of evolutionary psychology. He kindly looked at multiple drafts, providing much encouragement during the writing process. Without his initial suggestion and unfailing support, this book might not have been written.

My thanks also to the commissioning and editorial staff at Palgrave Macmillan, especially to Brendan George, Senior Commissioning Editor (Philosophy), and Esme Chapman, Assistant Editor (Philosophy), who have been thoroughly supportive of this project right from the start.

Finally, my deepest thanks to my family, for their patience, their encouragement and their support throughout this endeavour.

Introduction

What today we call 'evolutionary psychology' was pioneered by John Tooby and Leda Cosmides in the early 1990s and was subsequently developed and popularised by Steven Pinker, David Buss and Jerome Barkow among others. To its champions, evolutionary psychology is not just another research programme in the behavioural sciences: it is a revolutionary paradigm of the behavioural sciences. Evolutionary psychology is typically understood in terms of strong commitments: to particular views about evolution, to a particular view about the mind, as being an explanatory project, as being a metatheory for psychology in particular and for the behavioural sciences more broadly. Evolutionary psychologists claim to have shed new light on our psychology and to be working towards making visible the hidden processes of the mind. Evolutionary psychologists have also claimed that those insights have important public policy implications.

Great claims make for great controversy. And this has not been in short supply. Evolutionary psychology is unceasingly subject to a firestorm of criticism and controversy from both philosophers and social scientists. Articles and books carry the titles 'Pop Sociobiology Reborn' (Kitcher and Vickers, 2003), 'Perverse Engineering' (Haufe, 2008), *Getting Darwin Wrong: Why Evolutionary Psychology Won't Work* (Wallace, 2010), and 'The Seven Sins of Evolutionary Psychology' (Panksepp and Panksepp, 2000). If these titles are insufficient to convey the flavour of the sceptical viewpoint, one might cite the closing of Richardson (2007: 183): evolutionary psychology is to be dismissed as 'idle Darwinizing'. Buller (2005: 481) dismisses evolutionary psychology as being 'wrong in almost every detail'. And Hamilton (2008: 105; not to be confused with W. D. Hamilton) is on particularly dramatic form: 'evolutionary psychology is empirically unwarranted and conceptually incoherent to such an extent that it is a matter of professional sociological concern why it has come to achieve such a degree of popularity'. Sceptics paint a grim portrait of a research programme characterised by intellectual thinness, outdatedness and methodological inertia. If one were writing a script for a bad science or a pseudoscience, all the elements would be in place.

There is much to support and recommend the sceptical case. For all the fanfare and the fireworks, for all the theoretical posturing and promises, evolutionary psychology explanations are shockingly naked in historical and scientific detail—the kind of detail required to vindicate them. The jibe of 'just so' seems unshakable. Evolutionary psychology explanations have even been condemned on the basis being just so stories of the worst sort: of recasting banalities and stereotypes into an evolutionary mould. Rather than be a mirror of the past, they seem to be a mirror of our prejudices. Indeed, evolutionary psychology seems to always be but months away from yet another damaging episode of controversy and angry claims and counterclaims.

As one might expect, all of this forms a powerful impression that, no matter its output, evolutionary psychology is discredited, something one can legitimately dismiss wholesale. But I'm not so sure. I have come to believe that there is profound disconnect between discussions in the literature about what evolutionary psychology is presented as being and the realities on the ground about what it is actually doing. I believe that not only may the champions of evolutionary psychology have fallen for the marketing hype, but so too have many of their sceptics. The excess packaging and promotion of evolutionary psychology has comprehensively overshadowed what it's actually about. The reality lies somewhere else—less dramatic and less exciting in some ways but more dramatic and more exciting in other ways, ways largely invisible in the discussion but visible in the day-to-day activities of its practitioners.

Rather than thinking of evolutionary psychology as a paradigm, as a pretender metatheory of the evolutionary behavioural sciences, evolutionary psychology is better thought of as a research programme within the evolutionary behavioural sciences. Evolutionary psychology is a hypothesis-driven empirical science. The daily practice of evolutionary psychology is to focus on adaptive problems, hypothesise psychological adaptive solutions and to subject these hypotheses to testing. Its day-to-day activities simply lack the characteristics and credentials of anything remotely like a paradigm.

And rather than thinking of the programme as being explanatorily driven, of being motivated to simply explain observed phenomena in an adaptationist way, it is better thought of as heuristically driven, of being motivated to investigate whether psychological traits have features to be expected if they are solutions to adaptive problems, features hitherto unknown, and to investigate the possibility of hitherto unknown psychological traits on the basis of adaptationist considerations.

Heuristics are a largely untapped way of elucidating evolutionary psychology. Heuristics can be thought of as occupying the Goldilock zone of scientific discovery. They occupy the warm space between the bright star of algorithm, a process that guarantees a solution, and the infinite empty space of blind trial and error. They represent a constrained yet creative habitat, where new discoveries are slowly nurtured and grown, and blossom. Elucidating evolutionary psychology as a heuristic illuminates those aspects of adaptationist methodology in psychology that deserve serious attention. It reveals a unique set of concepts, research strategies and recipes for generating new and bold discoveries in psychology—a searchlight that can identify and chart plausible and unexplored research trajectories, a constrained yet creative way to win new facts that would otherwise be lost to the infinite space of random trial and error.

Furthermore, not only is evolutionary psychology better thought of as a research programme of heuristic exploration and discovery within the wider evolutionary behavioural science, its tenets ought to be unbundled and reframed only according to what its best practice requires. Accordingly, I want to identify, champion and vindicate a streamlined evolutionary psychology, one that navigates away from unnecessary controversy, one that better reflects the actual practice of evolutionary psychology on the ground, and one that doesn't overshadow what's valuable about the programme. It's both a conservative and a radical move to make. I think it's even an exciting move to make, for it seeks to focus attention on what's really worth fighting for in evolutionary psychology. I believe it's the sort of argument concerning evolutionary psychology that needs to be made, the sort of argument that is overdue, and the way I present it should make it clear that it's an attempt to avoid the theoretical and critical excesses on both sides of the debate.

Reading this far, you might already be realising the benefits that an unbundling and reframing of evolutionary psychology's framework can offer. To streamline the framework is to streamline the points of contention, liberating the programme from the suffocating shackles of controversy. To streamline the framework is to make visible what's valuable about the programme. To streamline the framework is to make its insights available to a wider constituency.

So the argument that I am making in *Rethinking Evolutionary Psychology* represents something unique in the literature: it recognises and criticises the excessive packaging, indeed marketing, of evolutionary psychology as a 'paradigm', one that can 'unify the social sciences', one that has 'public policy implications' and can 'reveal our place in the grand scheme of the natural world'. It recognises many of the deeply felt methodological worries about the standard depiction of the programme. And yet is also recognises that beyond the theoretical and theatrical hype, evolutionary psychology has generated an impressive portfolio of successful novel predictions—and seeks to unbundle and reframe both evolutionary psychology and the debates surrounding it on the basis of that programme, thereby honouring yet redirecting the established criticism, releasing the programme from much of the controversy, making better sense of how the programme fits within the wider evolutionary behavioural sciences and helping to make possible the kind of multi-disciplinary activities ultimately needed to deepen adaptationist hypotheses and explanations.

This book will cover a lot of ground, will make a number of twists and turns, and so it is expedient to provide a brief roadmap of our journey ahead. The first chapter, the shortest of the book, motivates and explicates evolutionary psychology as a paradigm. We start by highlighting the fundamental reasonableness of evolutionary psychology. The daily practice of evolutionary psychology is to focus on adaptive problems, hypothesise adaptive solutions and then subject these hypotheses to testing. Sounds reasonable enough. Why, then, the controversy? To begin to understand the controversy is to identify and to understand the tenets commonly associated with evolutionary psychology: empirical adaptationism; a strong form of methodological adaptationism; usually a strong form of massive modularity; inferring strong massive modularity from empirical adaptationism; a unification agenda for the behavioural sciences; and a public policy agenda. This will set the stage for the remaining chapters.

The second chapter explicates the case against evolutionary psychology as a paradigm. Not all that glitters is gold and a number of common criticisms don't succeed in their objective in virtue of being founded on misconceptions. Indeed, the literature is frequently disfigured by misconceptions about basic evolutionary psychology concepts. These mistakes resemble the classic whac-a-mole game: no matter how many times a mistake is hit over the head and corrected, it's only a matter of time before it reappears. Some of these mistakes are sufficiently common and important to be worth correcting, especially as they pertain to concepts that will be drawn on later in the book. Evolutionary psychology is often characterised as: (i) being conceptually outdated in virtue of multi-level selection; (ii) proposing that we're the Flintstones in modern attire, struggling to cope with this post-industrial world; (iii) proposing that psychological adaptations are rigid; (iv) consenting to, or requiring, a discredited picture of development; and (v) proposing nice-and-tidy, encapsulated psychological adaptations. The reality, as we shall see, is rather different.

More powerful criticism focuses on the general inference from empirical adaptationism to massive modularity and, most dangerously, the particular inferences used in methodological adaptationism to generate hypotheses. For example, there is an apparent lack of discipline in generating evolutionary psychology explanations. As noted by Gray et al. (2003), Rosen (1982) jibed that there are only two limiting factors that constrain adaptive explanations: the imagination of the theorist and the gullibility of the audience. Another line of scepticism, most recently updated by Richardson (2007), goes like this: when framed as explanations, evolutionary psychology research is remarkably sketchy in historical details. Such explanations are incomplete. And they're likely to remain incomplete. From this, Richardson and others dismiss evolutionary psychology as speculation, something safely to be ignored. Going through these and other adjacent criticisms, the case against evolutionary psychology will increasingly appear devastating. Case closed: that, indeed, is the near-settled judgement in philosophical treatments of evolutionary psychology.

The third chapter represents a dramatic change at how we look at evolutionary psychology. Despite its impressive force, the sceptical case sits awkwardly with the fact that evolutionary psychology has generated successful novel predictions. This situation—the stark contrast between philosophical dismissal of evolutionary psychology and predictive success of evolutionary psychology—should strike reasonable people as perplexing and unsatisfactory. After all, how can something totally without merit produce successful novel predictions? A portfolio of successful novel predictions signals that there is something of value operating. It signals the need to take a closer look at evolutionary psychology and to make sense of and articulate this value. It demands further investigation.

In this chapter, we will look at evolutionary psychology's heuristics. We shall see that the underlying logic of evolutionary psychology discovery is this: *if* psychological trait T is an adaptation for X, then psychological trait T should have configuration C, and so we should find phenomenon P. That is, if T is an adaptation for X, we should see signatures of this in the adaptation's design, in the kinds of inputs it receives, in the kinds of output it produces and so on, and this can bring into sharp view behavioural patterns we've never observed before. More precisely, evolutionary psychology can empower researchers to (i) discover new design features of extant psychological traits and (ii) discover hitherto unknown psychological traits.

The fourth chapter unbundles and reframes evolutionary psychology's theoretical framework. What's the minimum we can work with? What tenets do evolutionary psychologists actually need to subscribe to in order to pursue adaptationist hypothesising in psychology? What is striking is how few of the tenets commonly associated with evolutionary psychology, some of which are greatly contentious, have anything to do with evolutionary psychology as practised. If evolutionary psychology is about discovering psychological adaptations, why package all these tenets together? It's time to prize apart the tenets associated with evolutionary psychology. Understanding evolutionary psychology as a programme of heuristic exploration and discovery, rather than as a paradigm, happily liberates the programme from much of the criticism aimed against it. For example, by prising away tenets such as a commitment to massive modularity we prise away all the controversy that tenet generates. Ditto the other tenets. Interestingly, it also disarms many of the explanatory criticisms—criticism that is the source of much unease and mistrust about the programme. For example, failure to establish all evidential particulars, either at the present moment or in the final verdict, of an evolutionary psychology explanation does not in any way undermine the utility and value of adaptationist hypothesising. Indeed, understanding evolutionary psychology as a programme of discovery enables us to locate its unique place and role in the evolutionary and behavioural sciences. Establishing the multiple lines of evidence needed to fully vindicate adaptationist explanations requires multidisciplinary activity. In virtue of their successful novel predictions, evolutionary psychologists can help stimulate multidisciplinary activity. So while it's reasonable for sceptics to demand greater evidence for evolutionary psychology explanations, they're wrong in demanding evolutionary psychology alone satisfy such demands—these demands are properly allocated to the evolutionary and behavioural sciences collectively.

The fifth chapter seeks to restructure the debate. If evolutionary psychology is to realise its potential of stimulating multidisciplinary activity towards establishing the multiple lines of evidence needed to fully develop and vindicate adaptationist explanations, the evolutionary psychology debate needs depolarising. In virtue of the preceding chapters, for the first time the interpretive and strategic mistakes of the two opposing parties are visible. I argue that the strong sceptics' basic interpretive mistake is to overlook the heuristics of the programme and to instead overly focus on the current evidential status of psychological adaptation claims. This, in turn, creates a cascade of misunderstandings. I also advocate that leading evolutionary psychologists recognise their strategic mistake in obscuring or overshadowing the heuristics by packaging the programme with tenets irrelevant to its practice. Indeed, even naming their programme 'evolutionary psychology' appears to have been a fundamental strategic mistake, for it generates an incorrect but powerful expectation of what the programme is actually about. It creates cognitive mischief. Finally, a new and (in some respects) more moderate but equally subversive scepticism will also need to be identified and addressed—subversive because this strand of scepticism 'requires' that evolutionary psychology 'change its daily practice'—and so this strand arrives at the same endpoint as the more popular and established hard scepticism. This line of thought recognises that evolutionary psychology has been predictively useful but because there is 'new thinking' and developments in evolutionary theory, evolutionary psychology should update and do radically different things other than just adaptationist hypothesising. The notion that evolutionary psychologists should be doing things other than generating testable hypotheses will be strongly contested. Evolutionary psychologists have a unique role to play in the evolutionary behavioural sciences. They have a specialised job to do. If they didn't perform this role, the evolutionary behavioural sciences would be impoverished. Yes there is more to evolutionary theory than adaptationism and more methodological tools available than just adaptive thinking—but that speaks in favour of new programmes in the evolutionary behavioural sciences, of new research programmes to operationalise new thinking and new developments in evolutionary theory, not for the methodological adaptationism to be retired.

The book ends by advocating that strong sceptics, prominent champions of the programme, and the milder sceptics focus on, refocus on and revalue the heuristics of the programme respectively. As evolutionary psychology is in the business of generating and testing hypotheses about psychological mechanisms, then it is not only eminently reasonable but it should be welcomed by all. Only the most contentious of spirits will still find reason to fault its streamlined theoretical and methodological foundations. If the various parties endorse this picture, this streamlined evolutionary psychology, the polarisation can end and the fruits of genuine multidisciplinary research on purported psychological adaptations can begin to grow.

Philosophy has a special role to play in fostering greater understanding of arguably the most prominent and active application of evolutionary theorising in the behavioural sciences. Because of the typical way it has been packaged and promoted, many people, including philosophers have generated a flawed picture of evolutionary psychology, which leads to their extreme dismissal of the programme. *Rethinking Evolutionary Psychology* offers an alternative picture, a picture that better reflects its actual practice, a picture I think many can endorse, as well as one that conveys the vibrancy and possibilities of this type of reasoning—and its legitimate and important place in the evolutionary and behavioural sciences.

1
Evolutionary Psychology as a Paradigm

1.0 Introduction

We have evolved by natural selection. Even if not everything about an organism is an adaptation for survival and reproduction, it seems inescapable that organisms are well adapted, at least to the environment in which they evolved. Most consent to that proposition. But there is disagreement, often quite animated disagreement, as to what implications this has for psychology.

Our ancestors faced a multitude of recurrent survival and reproduction problems. In order to survive and reproduce, one must locate and secure resources; avoid pathogens; avoid predators; find a mate; counter threats from mate poachers; form and maintain coalitions and relationships; establish and protect one's status in a societal hierarchy; and so on. These problems are known as 'adaptive problems' and solutions to these problems promote reproductive success.

One proposal, perhaps the dominant view, is that selection favoured one or a few general-purpose psychological mechanisms, general learning and decision-making mechanisms sufficient to generate multiple behavioural solutions to multiple adaptive problems. If this is the case, applying evolutionary perspectives to psychology might have limited mileage.

An alternative proposal is that some or many of the problems were solved by dedicated psychological adaptations. Just as we inherit physiological specialisations, so too we might inherit psychological specialisations. If this is the case, this opens up the possibility that evolutionary theory can open up new lines of research in psychology. This is the possibility evolutionary psychologists pursue.

Quite reasonable, isn't it? If selection pressures have favoured physiological adaptations, then they might also favour psychological adaptations, and that this perspective might be useful in understanding psychology in some way. In a way, pursuit of this line of thought is 'inevitable' (de Waal, 2001). Of course, whether this kind of thinking is actually useful needs to be established.

Yet despite its apparent reasonableness, evolutionary psychology has been subject to a firestorm of criticism and controversy, often from philosophers of science. Dawkins observes that evolutionary psychology is 'subject to a level of implacable hostility, which seems far out of proportion to anything sober reason or even common politeness might sanction' (2005: 975). This book's introduction already observed some flashes of this deeply motivated hostility and as we go along we shall witness similar expressions.

So we have something of a puzzle: evolutionary psychology's underlying impulse—to hypothesise possible adaptive problems and to hypothesise possible psychological adaptive solutions—seems reasonable and yet evolutionary psychology is widely dismissed wholesale in philosophy and by many practitioners in the social sciences.

To begin to understand this puzzle, this controversy, is to understand that evolutionary psychology is frequently presented as much more than simply operationalising a set of ideas to illuminate our understanding of psychology. In introductory paragraphs and articles, theoretical manifestos and introductions to edited volumes, evolutionary psychology is presented as something of a game-changer. Something revolutionary. More precisely, it is frequently characterised in terms of commitments to particular views about evolution, to a particular view about the mind, as being an explanatory project, as being a metatheory for psychology in particular and the behavioural sciences more broadly, and as having a public policy agenda.

More than a research programme in the evolutionary behavioural sciences going about its daily business, it is championed as a 'scientific revolution' in psychology and the social sciences—as an all-encompassing view of science and, indeed, the world. We are assured that evolutionary psychology is 'becoming the unifying paradigm upon which the entire field of psychology can be based' (Fitzgerald and Whitaker, 2010: 284). 'Evolutionary psychology represents a true scientific revolution, a profound paradigm shift in the field of psychology' (Buss, 2005a: xxiv; see also Buss, 1995). And as you might already be realising, such a packaging of strong views is likely to magnetise not only supporters, but also opponents.

To understand how the underling impulse of evolutionary psychology, its heartbeat, is cashed out into a paradigm is to understand what its main theoretical and methodological tenets are commonly purported to be. This is the task we shall undertake in this short chapter, which sets the stage for the remaining chapters and forms a roadmap for the controversies that lay ahead.

I interpret the leading proponents of evolutionary psychology—Leda Cosmides, John Tooby, David Buss and others—as presenting evolutionary psychology as a package of views:

(T1) Evolutionary theory, with an emphasis on natural selection and adaptation
(T2) The possibility of psychological adaptations
(T3) Empirical adaptationism
(T4) Inference from empirical adaptationism to massive modularity
(T5) Methodological adaptationism
(T6) Metatheory for psychology and the behavioural sciences
(T7) Public policy agenda

The distinction between empirical and methodological adaptationism is owing to Godfrey-Smith (2001). Godfrey-Smith discusses these distinctions with respect to evolutionary biology but they're equally applicable to evolutionary psychology. Leading evolutionary psychologists seem to have no trouble subscribing to both empirical and methodological adaptationism, and there's plenty of evidence they do.

We'll go through each of these tenets in order. In doing so, we will not only be able to see precisely where and how the paradigmatic framework emerges, but also to see clearly and to anticipate precisely at which junctures controversy and heated resistance to the paradigm can arise. This is important: in the sceptical literature one is often treated to a battery of objections against evolutionary psychology, and without knowing the sequence of tenets that form the evolutionary psychology paradigm and how they relate to one another it can be unclear just how destabilising any particular objection is. And given sceptical writings often codify a number of objections against evolutionary psychology, this can, quite naturally, lead to a wholesale dismissal of evolutionary psychology—even if the objections tabled might not, in fact, justify such a judgement.

1.1 Natural Selection and Adaptation

The modern evolutionary synthesis, the synthesis of Darwin's theory of evolution by natural selection and Mendelian inheritance, forms the basic theoretical foundation of evolutionary psychology. This synthesis is currently the consensus framework of modern evolutionary biology. The modern evolutionary synthesis has many components and concepts, which are well known and can be found in many textbooks. Instead of repeating them, I'll only identify the concepts most directly relevant to evolutionary psychology.

1.1.1 Evolution by Natural Selection

Biological evolution is a change in the characteristics of a population over time. More precisely, a standard definition is that evolution occurs precisely when there is a change in the gene frequencies found in a population (Sober, 2000). Populations evolve, not individuals—individuals survive and reproduce. Several processes drive evolutionary change: natural selection (heritable variation in fitness), genetic drift (gene frequencies can change owing to chance alone), migration (gene flow) and mutation. In the classical view, within a population, selection and drift decrease genetic variation, while mutation and migration increase genetic variation (Culver, 2009).

Evolutionary psychologists focus on natural selection. Evolution by natural selection is the process of preserving and increasing the frequency of fitness-enhancing characteristics or traits in a population. For selection to occur in a population, three conditions must obtain. First, variation: individuals in a population must differ with respect to the relevant trait in question. Without variation, all individuals will have the same trait value, and will thereby be indistinguishable with respect to that trait. Second, heritability: the variation found in the population must be heritable to some degree. Non-heritable traits, and the advantages they provide, cannot be passed on from one generation to the next. Third, differential fitness: individuals must have a probability of reproduction that is a function of the trait value in question. Although this classic formulation presents selection as acting on individual organisms, one can view selection as acting on genes, groups or species.

1.1.2 Adaptations and By-products

Natural selection results in the evolution of adaptations, where selection increases the frequency of beneficial alleles in a population to the point of fixation. An adaptation can be defined as an inherited and reliably developing trait that was selected for because it helped solve an adaptive problem—a recurring problem of survival or reproduction (Buss, 2008; see also Stearns 1986 and Tooby and Cosmides 1992). More precisely,

> Characteristic *c* is an adaptation for doing task *t* in a population if and only if members of the population now have *c* because, ancestrally, there was selection for having *c* and *c* conferred a fitness advantage because it performed task *t* (Sober, 2000: 85).

This way of defining an adaptation—adaptation as an historical concept—represents a 'reasonable consensus on what it is to be an adaptation' (Godfrey-Smith, 1998: 191). This is also how the concept is commonly defined in evolutionary psychology. Accordingly, this book adopts that definition. For a discussion of other ways of defining adaptation, see Shanahan (2004).

Adaptations are to be distinguished from their by-products. A by-product is a trait that evolved not because it was selected for, but because it was connected to another trait that was selected for. By-products do not function to solve adaptive problems: the sound of the heart pumping blood and the white colour of bones are non-functional by-products of selection.

1.1.3 Adaptations Advantageous on Balance

To be selected for a trait need not be advantageous under every scenario. Rather, all that is required is that the trait is advantageous overall. For example, a child's enlarged skull displays signs of being an adaptation, as a larger brain is correlated with greater cognitive power. But this makes birth complicated. And dangerous: there's a risk of childbirth death. Yet the enlarged skull is advantageous on balance, and so could have been selected for.

1.1.4 Adaptations and Constraints

Natural selection optimises under constraints. 'Adaptationist models do predict optimality, but this optimality is always *constrained* optimality' (Sansom, 2003: 497; original emphasis). Genetic and developmental constraints can prevent certain traits and designs being realisable. Coordination constraints with extant mechanisms can prevent the evolution of certain traits and designs. Prohibitive costs can also rule out the evolution of certain traits and designs. Furthermore, 'Local optima can prevent the evolution of better adaptive solutions that might, in principle, exist in potential design space' (Buss *et al.*, 1998: 538). In other words, the required fitness chain between one solution and a better solution might not obtain. Buss *et al.* (1998) invite us to think of selection as building adaptations through a relentless mountain-climbing process. On top of a neighbouring mountain a better design might be found 'but selection cannot reach it if it has to go through a deep fitness valley to get there' (*ibid.*: 538).

1.1.5 Adaptation/Adaptive Distinction

We need to distinguish between a trait being an adaptation and a trait being adaptive.

> An *adaptation* is a character favoured by natural selection for its effectiveness in a particular role; that is, it has an evolutionary history of selection. To be labelled as *adaptive*, a character has to function currently to increase reproductive success (Laland and Brown, 2002: 132, original emphasis).

Hence, 'To say that a trait is an adaptation is to make a claim about the cause of its presence; to say that it is adaptive is to comment on its consequences for survival and reproduction' (Sober, 1993: 211). It follows that a trait can be an adaptation without being adaptive, and a trait can be adaptive without being an adaptation.

The adaptation/adaptive distinction highlights the possibility of a mismatch between adaptations and novel environments. Evolutionary psychologists stress that our contemporary environment differs from our ancestral environment. When an organism's environment changes speedily and significantly, behaviours that were adaptive in the previous environment can become dysfunctional or deleterious in the new environment, undermining reproductive success. Some evolutionary psychologists call such behaviour 'maladaptive' (e.g. Tooby and Cosmides, 2005). Sterelny (1995) adopts and illustrates the distinction between adaptive and maladaptive behaviours with the memorable example of hedgehogs responding to danger by rolling into balls: an effective response to dangers from its natural predators, but a poor response to the danger of cars. More on this in the next chapter.

1.2 The Possibility of Psychological Adaptations

It's common to think of the outcomes of selection as being exclusively physiological. Many physiological traits—hearts, livers, lungs and so on—are widely seen as having been shaped by natural selection for the survival and reproduction of their possessors. But why shouldn't the same be true of psychological traits? As Dawkins notes, the claim that psychology is on the same footing as the body where selection is concerned is 'exceedingly modest' (2005: 978).

A guiding principle of evolutionary psychology, what I call its symmetry principle, is that there is no reason to suppose that only physiological traits can be selected for. This enables us to move from physiological adaptations to psychological adaptations. As with physiology, so too with psychology: 'Just as a shared set of digestive mechanisms both enable and constrain the diverse diets of human populations, so do a comparable set of behavioral mechanisms enable and constrain our social-cultural behavior' (Barkow, 2006: 21–2).

Physiologists have divided the body up into different physical organs, based largely on the purported functions of organ. As with physiology, so too with psychology: we can do this with psychology, too, with psychological 'organs'. As Cosmides and Tooby (1997a) put it,

> Our body is divided into organs, like the heart and the liver, for exactly this reason. Pumping blood through the body and detoxifying poisons are two very different problems. Consequently, your body has a different machine for solving each of them. The design of the heart is specialized for pumping blood; the design of the liver is specialized for detoxifying poisons. Your liver can't function as a pump, and your heart isn't any good at detoxifying poisons. For the same reason, our minds consist of a large number of circuits that are functionally specialized.

Hence, we can hypothesise that a particular psychological trait exists in the form that it does because it solved a specific problem of survival or reproduction recurrent in ancestral populations. For example, we can hypothesise that fear of snakes is an adaptation, taking the form that it does because it solved a specific and recurring problem in ancestral populations, namely the threat of being poisoned by snake bites (Öhman and Mineka, 2001). As those who possessed this trait would enjoy a fitness advantage over those who didn't possess the trait, the trait would be selected for and become prevalent in ancestral populations.

As our physiological adaptations are species-typical, appearing in normally developing members of the human species, our psychological adaptations are also held to be species-typical, although, of course, they can be sex- and age-specific (Buss, 2008). And as physiological adaptations tend to be problem-specific (hearts, lungs and livers, etc. function to solve specific problems), psychological adaptations are also held to be problem-specific—also known as 'domain-specific' (Goetz et al., 2009).

Psychological adaptations, also known as 'evolved psychological mechanisms', are understood to be functionally specialised mechanisms: 'complex structures that are functionally organised for processing information' (Tooby and Cosmides, 1992: 33). This reflects evolutionary psychologists' information processing view of psychology.

Buss (1995) provides what I judge to be the clearest and most decisive definition of what a psychological adaptation is. According to Buss, a psychological adaptation is a set of processes that:

> 1. Exists in the form it does because it (or other mechanisms that reliably produce it) solved a specific problem of individual survival or reproduction recurrently over human evolutionary history.
>
> 2. Takes only certain classes of information or input, where input (a) can be either external or internal, (b) can be actively extracted from the environment or passively received from the environment, and (c) specifies to the organism the particular adaptive problem it is facing.
>
> 3. Transforms that information into output through a procedure (e.g., decision rule) in which output (a) regulates physiological activity, provides information to other psychological mechanisms, or produces manifest action and (b) solves a particular adaptive problem (1995: 5–6).

So psychological adaptations solved specific problems in ancestral environments, are triggered by only a narrow range of information, and are characterised by a particular set of procedures or decision rules, and produce behavioural output that solved the adaptive problem in ancestral times. In other words, they are 'functionally specialised' or 'modular'. More on this in the next chapter.

Postulating psychological adaptations as information processing mechanisms should not be taken to mean that the outputs of these mechanisms are bereft of emotion. Quite the contrary: evolutionary psychologists stress that such output is often experienced as emotional states (Cosmides and Tooby, 2000). Indeed, emotional states are clearly useful in helping solve adaptive problems: to prioritise certain behaviours; to shift attention to potential threats; to be attracted to potential mates; and so on.

As Griffiths (2007: 407) points out, when evolutionary psychologists 'present experimental evidence of domain specificity in cognition, it is generally evidence suggesting that information about one class of stimuli is processed differently from information about another class of stimuli—that is, evidence of the use of different proprietary algorithms in the two domains'. So, to cite an example that we will look at in a moment, Cosmides and Tooby interpret various Wason selection tasks as demonstrating that certain ways of describing a task can activate a domain-specific mechanism for detecting cheaters.

Finally, it's important to understand that we're often unaware of the existence and operation of these information processing mechanisms. For example, when people venture out to pursue short-term sexual encounters, clearly they're not consciously calculating levels of heritable fitness and the extent to which this might translate into fitness advantages for potential offspring. Rather, they experience sexual desire and attraction. The significance of this point is often underappreciated, a point we'll discuss in the final chapter.

1.3 Empirical Adaptationism

How powerful and widespread is natural selection as an evolutionary process? Empirical adaptationism is the position that selection is powerful and widespread. Godfrey-Smith (2001: 336) defines empirical adaptationism as follows:

> Empirical Adaptationism: Natural selection is a powerful and ubiquitous force, and there are few constraints on the biological variation that fuels it. To a large degree, it is possible to predict and explain the outcome of evolutionary processes by attending only to the role played by selection. No other evolutionary factor has this degree of causal importance.

This is the usual understanding, probably the default understanding, of the term 'adaptationism'. Prominent evolutionary psychologists frequently consent to empirical adaptationism. For example, Tooby and Cosmides (1992: 52) claim,

> To the extent that a feature has a significant effect on reproduction, selection will act on it. For this reason, important and consequential aspects of organism architectures are shaped by selection. By the same token, those modifications that are so minor that their consequences are negligible on reproduction are invisible to selection and, therefore, are not organized by it. Thus, chance properties drift through the standard designs of species in a random way, unable to account for complex organized design and, correspondingly, are usually peripheralized into those aspects that do not make a significant impact on the functional operation of the system.

It's useful to distinguish between the power of selection and the ubiquity of selection. Sober characterizes (empirical) adaptationism in terms of various hypotheses. Given a population *X* and a trait *T*, Sober (2000: 124) distinguishes between the following hypotheses:

> (U) Natural selection played some role in the evolution of *T* in the lineage leading to *X*.
>
> (I) Natural selection was an important cause of the evolution of *T* in the lineage leading to *X*.
>
> (O) Natural selection was the only important cause of the evolution of *T* in the lineage leading to *X*.

These hypotheses are ranked according to their strength: (O) is the strongest hypothesis and implies (I) and (U); (I) is weaker than (O) and implies (U), and (U) is the weakest. By generalising (O), Sober (2000: 124) arrives at his formulation of (empirical) adaptationism as follows:

> [Empirical] Adaptationism: Most phenotypic traits in most populations can be explained by a model in which selection is described and nonselective processes are ignored.

This definition of (empirical) adaptationism holds that selection is the *most* important cause of *most* traits in *most* populations. Empirical adaptationism can be framed in a weaker way by, for example, claiming that selection is the most important cause of *some* traits in most populations. Likewise, it can be framed in a stronger way by claiming that selection is the most important cause of *all* traits in most populations. These different versions of empirical adaptationism can be tested in the long-run by proposing and testing various selectionist hypotheses. Empirical adaptationism's confirmation obviously increases as further adaptationist hypotheses are confirmed, and it obviously weakens as adaptationist hypotheses fail. Hence, empirical adaptationism is vindicated 'if in the majority of cases a better fit to the data is achieved by a selection-based model than is achieved by any other model of comparable complexity' (Godfrey-Smith, 2001: 344–5).

1.4 Inference from Empirical Adaptationism to Massive Modularity

From general considerations of empirical adaptationism, can we securely make any inferences about what kind of overall cognitive architecture is likely to have evolved? Many prominent evolutionary psychologists believe we can. They believe that from considerations of empirical adaptationism, we can securely infer that the mind is probably massively modular: a constellation of hundreds, possibly thousands, of functionally specialised information-processing adaptations. Particular adaptationist hypotheses have been formulated for face recognition; spatial relations; tool use; fear; social exchange; emotion perception; kin-orientated motivation; friendship; grammar; theory of mind; and many more (Tooby and Cosmides, 1992)

Note that we can distinguish between two senses of massive modularity: a moderate form, which claims that the number of psychological adaptations is significantly greater than has been traditionally believed, and the stronger form, which claims that there are no domain-general mechanisms ('domain-general' means the mechanism is not dedicated to a particular domain or problem). Note, the moderate form of massive modularity is not claiming that there are domain-general mechanisms—it's simply not ruling them out.

A key metaphor used to illustrate this massive modular state of affairs is the Swiss Army Knife. The Swiss Army Knife is multi-tooled, with sharp blades and other tools. The point of this metaphor is not that each psychological mechanism is physically separate from other mechanisms—indeed, as we shall see in the next chapter, how a psychological mechanism is instantiated can be physically quite messy. The point of the metaphor is that each blade and tool is specialised. Each blade and tool is designed for a specific task or function: the blade for cutting; the screwdriver for removing screws; the corkscrew for drawing out corks. A screwdriver would be ineffective in drawing out corks, more likely mashing the cork up than cleanly removing it. A corkscrew would be of little use removing screws. And likewise our psychological mechanisms are designed for a specific task or function: a mechanism specialising in solving the adaptive problem of infidelity will be of little use when solving the adaptive problem of foraging.

1.5 Methodological Adaptationism

Another type of adaptationism Godfrey-Smith recognises is methodological adaptationism.

> Methodological Adaptationism: The best way for scientists to approach biological systems is to look for features of adaptation and good design. Adaptation is a good 'organizing concept' for evolutionary research (2001: 337).

Methodological adaptationism is a recommendation for how to go about investigating nature—specifically, to propose and pursue adaptationist hypotheses. Notice this definition slides together two distinguishable versions—the 'best way' of doing research and a 'good' way of doing research. Contextualised into psychology, they are:

> Strong methodological psychological adaptationism: the best way to approach psychological traits is to look for features of adaptation and good design. Adaptation is the best organising concept for psychological research.

> Moderate methodological psychological adaptationism: a useful or good way to approach psychological traits is to look for features of adaptation and good design.

Note, the moderate form is not trivial. It is not at all a given—it needs to be justified. And as we'll see later, even the moderate form has been challenged.

The main exponents of evolutionary psychology typically advocate the stronger form of methodological adaptationism. For example, Buss (1995: 6) holds that 'a central premise of evolutionary psychology is that the main nonarbitrary way to identify, describe, and understand psychological mechanisms is to articulate their functions—the specific adaptive problems they were designed to solve'. The main non-arbitrary way of individuating parts of the body is by reference to their evolved function—the heart is distinct from the liver, because the heart functions to pump blood, whereas the liver functions to detoxify. Likewise, Buss claims, evolutionary psychology provides the main non-arbitrary means of individuating psychological processes, by reference to their evolved function. Cosmides and Tooby (1997a: 14) also make this point, writing that 'knowledge of adaptive function is necessary for carving nature at the joints'.

How does evolutionary psychology methodology produce hypotheses? The hallmark of the methodology is reasoning between hypothesised adaptive problems and hypothesised adaptive solutions. As one can expect, this reasoning can start either from the position of the purported adaptive problem or from the position of the purported adaptive solution. For now, we can call this 'forward adaptationist reasoning' and 'reverse adaptationist reasoning' respectively.

If we are starting from the position of a purported adaptive solution, we first identify a psychological trait that has the hallmarks of being an adaptation—such as being developmentally robust and obtaining cross-culturally. We then reason backwards, from present to our ancestral past; we reason how it could have been selectively advantageous to have the trait, what kind of adaptive problem the trait could have solved.

If we are starting from the position of a purported adaptive solution, we first think about the kind of selective conditions and pressures that obtained in our ancestral past and from this identify a possible adaptive problem; we then consider what kind of psychological trait, if any, could solve the adaptive problem, what kind of trait would have been selectively advantageous to have and on the basis of this hypothesise that we may, indeed, have such a selected trait. Experiments are then designed to see whether we do have it.

Typically, reverse adaptationist reasoning is associated with explanation, with explaining extant behavioural phenomenon in adaptationist terms, while forward adaptationist reasoning is associated with discovery, with the discovery of previously unknown psychological traits. I believe the methodology is more nuanced and more sophisticated than that simple pairing. Indeed, the full power of the methodology will become visible only when the methodology is explicated in detail and in more precise ways. But I shall reserve explicating the methodology in detail until Chapter 3, when the dialectical pressure will benefit decisively from such a discussion.

An example of reverse adaptationist reasoning is enhanced altruism towards genetic offspring. It is a commonplace that people reliably show enhanced altruism towards their genetic offspring. Indeed, we tend to have a unique kind of love for our own related, genetic offspring. And after a child is born, generally, without conscious thought, without conscious calculation, and without hesitation, we pour our emotional and financial resources onto our genetic children. Not only that but we feel good for doing so. And we generally have these deep feelings and make these deep sacrifices for our own genetic offspring rather than non-genetic offspring. Of course, this is not always the case: parents have been known to reject even their genetic offspring. But as generalisations go, it's relatively uncontroversial.

How do we explain this state of affairs? What is the powerful motivation mechanism behind it? Why is it there? Note that preferential altruism towards our own genetic offspring appears to be developmentally robust and universal. Might it be an adaptation? Daly and Wilson (1980) believe so. They hypothesised that what they termed discriminative parental solicitude is an adaptation, an adaptive solution to the adaptive problem of preferentially directing altruism towards kin.

To see why this is an adaptive problem, consider that, in principle, altruism can be directed towards anyone. Suppose one's psychological constitution is such that one has strong feels for children irrespective of their genetic relatedness to oneself, that one has feelings for a non-genetic child with exactly the same degree of strength and quality as for a genetic child. Clearly, such indiscriminate behaviour will have significant fitness costs. Hence the need to channel such costly behaviour in ways that enhance fitness.

Given the problem of preferentially directing altruism towards kin, according to evolutionary psychology's conceptual scheme we should expect selection to solve this problem via a dedicated psychological mechanism, one that could reliably regulate parental solicitude, to ensure that parental care is focused only on those one is confident to be genetic offspring: a psychological adaptation for discriminatively allocating parental care to genetic offspring. Such a trait would have been selectively advantageous and so over vast evolutionary timescales refined and spread through the population.

This example will also be used to illustrate another methodological point: we can hypothesise behavioural phenomena not just in terms being generated directly by dedicated psychological mechanisms, but also in terms of being by-products of dedicated psychological mechanisms. If there is, indeed, a psychological adaptation for preferential altruism towards genetic offspring, then it's clear that such a mechanism will not be trigged by non-genetic individuals who come under our care, such as stepchildren. Accordingly, step-parents will tend to not feel solicitude for stepchildren. If this is indeed the case then abuse of stepchildren should be orders of magnitude more common than abuse of genetic children. In other words, step-parents should represent the single biggest risk factor for child abuse.

Here, child abuse is not being explained in virtue of a psychological adaptation for harm or infanticide. Rather, child abuse can be explained as a by-product of parental solicitude not being activated. And over two decades, Martin Daly and Margo Wilson have documented a large portfolio of evidence for the Cinderella Effect in both legal and non-lethal abuse. For example, they found that in England and Wales 103 children under the age of 5 were beaten to death by their stepfathers and 117 were beaten to death by their (presumptive) genetic fathers between 1977 and 1990. Given that less than 1% of children in England and Wales under the age of 5 lived with stepfathers and over 90% lived with their (presumptive) genetic fathers during the time-period, the difference in per capita rates of such lethal assaults of children under 5 is dramatic, being well over 100-fold (Daly and Wilson, 1994, 2007).

Let's now turn to forward adaptationist reasoning. The cheater detection hypothesis is often submitted as being the leading textbook example of a successful hypothesis generated by reasoning from the past to the present. Humans are social cooperators par excellence. But acts of cooperation carry with them risks of defection: one might perform an altruistic act only for the beneficiary to fail to reciprocate. A good solution to this adaptive problem would be the ability to detect cheaters, those who violate social contracts, and to reconfigure altruistic acts accordingly. On the basis of this adaptive problem, Cosmides and Tooby (1992) hypothesised that selection would favour the evolution of a mechanism dedicated to detecting, and avoiding future cooperative effort with, cheaters—a cheater detection mechanism, a psychological adaptation for representing and computing social exchanges. As the purported adaptation functions to reason effectively about social contracts, the mechanism would be activated only by content cues signally social exchange.

In order to vindicate this hypothesis, Cosmides and Tooby appealed to the Wason selection task (Wason, 1966), an experiment designed to investigate reasoning about conditionals. In the task, subjects are presented with a conditional rule, either in a descriptive form ('If something has property *P*, then it has property *Q*') or a deontic form ('If something has property *P*, then it should have property *Q*'). An example of the descriptive form of the Wason selection task is the card problem experiment. A rule is presented: 'if the letter "D" is on one side of a card, the number "3" is on the other'. Four double-sided cards are placed on a table. For each card, a letter appears on one side, a number on the other. Subjects can only see one side of each card: 'D', 'F', '3' and '7'. Subjects are invited to identify which of the four cards must be turned over in order to establish whether the rule is true or false. Subjects should identify *P* and *not-Q* cards as falsifying conditional statements of the form 'if *P*, then *Q*'. Hence, only the *P* card ('D') and the *not-Q* card ('7') needs to be turned over. Typically, performance on this version of the Wason selection task is poor: on average, only around 10% of subjects correctly identify that the *P* and *not-Q* cards are the only cards that need to be turned over.

However, subjects perform dramatically better in deontic forms of the Wason selection task. In Griggs and Cox (1982), for example, subjects were asked to imagine checking a bar to establish whether the following conditional rule was being obeyed: 'if a person is drinking beer, he or she must be over 20 years old'. Four cards conveyed information about people at the bar: 'beer', 'coke', '25 years old', '16 years old'. Typically, performance on this version of the Wason selection task is very good: on average, around 75% of subjects correctly choose the *P* card ('beer') and the *not-Q* card ('16 years old').

What are we to make of this? Cosmides and Tooby argue that the dramatic difference in performance between the card problem experiment and the bar problem experiment is owing to the latter experiment having social exchange cues that trigger the cheater detection mechanism to reason effectively over the conditional rule ('if you accept the benefit of drinking beer, you must satisfy the age requirement'), whereas the former experiment has no such cues, and so fails to trigger the operation of the algorithm.

1.6 Metatheory for Psychology

Prominent evolutionary psychologists frequently advocate that evolutionary psychology should act as a metatheory for not only psychology, but also the behavioural sciences and thereby radically reform the wider field. That's quite an ambition. Duntley and Buss (2008) are especially dramatic on this point. In the subsection 'What is Evolutionary Psychology?', they criticise a target article's summary of evolutionary psychology as failing to mention evolutionary psychology as a metatheory of psychology. Don't miss the final sentence:

> Evolutionary psychology provides a unifying metatheory for the currently disparate and disconnected branches of psychology (Buss, 1995). Evolutionary psychology unites the field of psychology with all the other life sciences, including biology, economics, political science, history, political science [sic], legal scholarship, and medicine; it unites humans with all other species, revealing our place in the grand scheme of the natural world (Duntley and Buss 2008: 31).

This ambition is motivated and fuelled by a deep dissatisfaction with the current state of psychology and the behavioural sciences more broadly. Indeed, Tooby and Cosmides explicitly indicate that the rationale that drove the founding of evolutionary psychology was 'the mutual incompatibility of models across the behavioural sciences, and their inconsistency with evolutionary biology' together with the ambition to unify the behavioural sciences within an evolutionary framework (2007: 42). Buss (1995) characterises psychology as being in conceptual disarray, with each division anxiously working on its own mini-theories, isolated from, and untouched by, empirical findings. Furthermore, Daly and Wilson observe that non-evolutionary theories in psychology 'have risen and fallen more like a succession of fashions than like the building blocks of a cumulative science' and that 'mainstream social psychology has gone in circles, such that work in the 1990s is in no clear sense an advance over that in the 1950s' (1999: 510).

So leading evolutionary psychologists see the existence of unanchored, unattached mini-theories to be problematic; they see their purported mutually incompatible to be objectionable; and they see the purported inconsistency of these mini-theories with evolutionary theory to be entirely unacceptable. And, as we just saw, proponents hold that evolutionary psychology provides the main non-arbitrary way of carving up the mind, providing the true psychology taxonomy; from this they infer that only evolutionary psychology can provide metatheory for psychology and the wider behavioural sciences.

1.7 Public Policy Agenda

Leading evolutionary psychologists also encourage evolutionary psychologists to consider, articulate and advocate public policy implications, if any, of their research. In the context of law, Cosmides and Tooby (2006: 182) frame the claim as follows: 'Which programs reliably develop in the human mind, and how do they process information? Evolutionary psychology seeks to answer this question. Accurate answers, when they are eventually arrived at, will have implications for lawmaking'.

And there is a point here: whether explicitly or (as is usually the case) implicitly, public policy is predicated on a certain picture of psychology, on what is possible and what is not possible. For thousands of years, stretching back to Plato, philosophers have conceptualised what many today still call 'human nature' and theorised political orders and institutions on the strength of such conceptualisations—recall the much repeated phrase 'Plato to Nato'. Much later, and more recently, social scientists have entered the picture, in particular conceptualising the mind in ways that presume relatively unconstrained behavioural flexibility with respect to conditioning, prize cultural learning exclusively, and explicitly or implicitly formulate and advance policy recommendations on that basis. Social scientists often do their work not only to understand the world, but also to change it. They often seek to know the causes of things in order to be able to change those things.

So when the new paradigm of evolutionary psychology was launched in the early 1990s, it seemed clear to some that a shift from an exclusively domain-general view to a massively modular view (typically in the strong sense but also advocated in the moderate sense), a shift from an exclusively cultural view of behaviour to one that emphasises an adapted psychology in a cultural setting, could lead to, perhaps even necessitate, a re-examination of public policy, perhaps even a radical overhaul of policy.

Evolutionary psychologists certainly have not been shy in charting possible implications of their hypotheses to public policy. In 1996, the policy think-tank *Demos* dedicated a quarterly issue to work connecting evolutionary psychology to public policy. Thornhill and Palmer (2000) dedicated several chapters looking at possible implications of their research on rape policy. In 2003, Albert Somit and Steven Peterson edited a volume on evolutionary psychology and public policy, a collection of papers exploring implications of evolutionary thought not only for domestic policy on prostitution, addictive behaviour, euthanasia, criminal justice policy, but also for international policy on agricultural biotechnology policy, ethnic conflict and state building, and international security. In 2004, Charles Crawford and Catherine Salmon also edited a volume on evolutionary psychology and public policy, a collection of papers on evolutionary psychology's purported applications to policy issues such as women in the workplace, marital support, child support and cultivating morality. More recently, Swami and Salem (2011) claim that evolutionary psychologists working on attractiveness have not paid much attention to practical applications of their research. They argue that this is a missed opportunity: evolutionary psychologists 'have a role to play in challenging the notion of ideal beauty and showing how and when beauty ideals can be challenged and changed' (*ibid.*: 163).

1.8 Conclusion

We started with a puzzle. Evolutionary psychology can be understood in a reasonable way: 'Evolutionary psychology is the attempt to understand our mental faculties in light of the evolutionary processes that shaped them' (Pinker, 1997a). And yet it can also be understood in an extraordinarily ambitious way: 'Evolutionary psychology is an attempt to unify the psychological, social, and behavioral sciences theoretically and empirically within a single, mutually consistent, seamless scientific framework' (Tooby and Cosmides 2008: 114). We are now in a position to make sense of this puzzle.

Selection solves adaptive problems. Some adaptive problems, like the pumping of blood, are solved by a physiologically specialised mechanism. Other adaptive problems seem to require psychological solutions—and there is no reason in principle why selection cannot shape cognitive systems and fashion them into functionally specialised behavioural mechanisms that solve those adaptive problems.

Where evolutionary psychology decisively moves beyond being simply a way of understanding our psychology to being a paradigm is in the following train of escalating of thought. If selection is, indeed, a powerful force, might it not have favoured an overall cognitive architecture massively populated by such mechanisms? Isn't that likely? Indeed, might it have favoured such a massive arrangement of modular mechanisms exclusively? Might it have actively selected against domain generality, eliminated domain-general intelligence? Indeed, is domain generality even possible in the first place? There are no domain-general problems—why then would there be domain-general solutions?

Riding this rapid train of thought has huge implications. According to evolutionary psychologists, much of the social sciences—the disunified carnival of middle and mini-theories inconsistent with one another and inconsistent with evolutionary theory—trade on the notion of domain-general intelligence and therefore appeal to a carnival of exclusively cultural processes to explain social phenomena. Hence, the train of thought that arrives at strong massive modularity has a huge implication: if selection really did favour strong massive modularity, not only would the notion of domain-general intelligence be displaced, but so also would the existing foundation of social science. Evolutionary psychology could form the new foundation of a new psychology and new behavioural science. This could be revolutionary. We could identify the confederation of functionally specialised systems in the mind, explain behavioural phenomena on this basis, unite and reorganise the arbitrary branches of psychology and the behavioural sciences into a seamless metatheory, and as social science seeks to inform policy, this new unified social science should lead to better public policy—no more flights of fancy about what is possible, but effective policy and effective outcomes.

These exciting possibilities were perhaps far too great a temptation to resist and when the research programme of hypothesising adaptive problems and adaptive solutions was launched it was immediately packaged within this ambitious paradigm. And that is precisely how it is usually pictured and understood:

> Evolutionary Psychology is a deeply ambitious enterprise. It presents us with a Grand Unified Theory of the structure and evolution of the human mind, and its proponents have busily gathered

> some provocative evidence from a number of intriguing hypotheses derived from that Grand Unified Theory—hypotheses about many of the most intimate aspects of our lives (Buller, 2005: 481).

Having seen and understood the key tenets of the evolutionary psychology paradigm and how they relate to one another, one might have the impression that evolutionary psychology has set itself up as a kind of city of light on the hill, a beacon for all psychologists and social scientists. The trouble with attempting to occupy such a high ground is that you will inevitably attract challenges. The higher up you establish yourself, the greater the scrutiny. Projectiles will be launched. Batteries will strike your walls and foundations. If there are vulnerabilities and weaknesses, they will be probed, exposed and exploited. In the next chapter we will see how the city on the hill has been fought against and, ultimately, broken through. Under relentless analysis, the series of theoretical moves charted in this chapter breaks down at crucial junctures.

2
Subverting the Paradigm

2.0 Introduction

In the previous chapter we saw how evolutionary psychology is characterised in terms of commitments to particular views about evolution, to a particular view about the mind, as being an explanatory project, as being a metatheory for psychology in particular and the behavioural sciences more broadly, and as having a public policy agenda. As one can expect, the more tenets are championed, the greater the range of possible criticism. Indeed, in the case of evolutionary psychology, practically every tenet has to some degree or another been objected to. And the fiercest sceptics see every tenet, every theoretical turn, as mistaken.

Undoubtedly, some of the objections are generated by conceptual misunderstandings. But others are serious and legitimate questions. Large-scale criticisms of evolutionary psychology are rarely unadulterated: works that deploy a wide range of arguments against evolutionary psychology tend to have a fair degree of conceptual misconceptions baked in. This can tempt those who are sympathetic to evolutionary psychology to dismiss scepticism wholesale in much the same way as the hard sceptics dismiss evolutionary psychology wholesale. I believe this to be unfortunate because those within the sceptical tradition, despite the occasional misconception, have legitimate and pressing questions and objections that need facing and answering. In this chapter, therefore, we shall seek to separate the chaff from the wheat, to identify precisely where criticism is overplayed and where criticism successfully subverts the paradigm.

One broad species of criticism is to claim that evolutionary psychology is inconsistent with something or is problematised by something: for example, with new thinking in evolutionary theory, multi-level selection, developmental considerations, behavioural flexibility, our high degree of adaptedness to our current environment and the possibility of recent selection in the past 10,000 years since the advent of agriculture. I believe much of this is overblown: certainly, developmental considerations, our relatively high degree of adaptedness to our current environment, behavioural flexibility and similar considerations do not destabilise or problematise evolutionary psychology as is commonly thought and the belief that they do seems to be due to common conceptual misunderstandings. As for what new thinking and developments in evolutionary theory like niche construction and gene–culture coevolution actually mean in the context of this debate, whether such developments mean (as some believe) that evolutionary psychology should change, I shall discuss in the last chapter because it will require us to first get a clearer and more secure picture and understanding of the role of evolutionary psychology in the evolutionary behavioural sciences.

The misconceptions that plague the sceptical literature are not only sufficiently common, and sufficiently important, to be worth correcting, but in doing so we also have a valuable opportunity to explicate in further detail key concepts that will be drawn upon in key ways in the next chapter. So in 2.1, I briefly look at the levels of selection debate and what impact, if any, this should have on evolutionary psychology. In 2.2, I identify the concept of evolutionary adaptedness, distinguish between the possibility of environmental mismatch and the ubiquity of environmental mismatch, and discuss whether the possibility of post-Pleistocene selection destabilises evolutionary psychology. In 2.3, I establish why psychological adaptations are often contextually flexible, and in 2.4 I set out why developmental robustness is consistent with developmental complexity and interactionism. And, in 2.5, I clarify what evolutionary psychologists mean when they employ the term 'module'.

Nevertheless, the case against evolutionary psychology cannot be fully reduced to misconceptions. As we shall see from 2.6 onwards to the end of the chapter, there are serious challenges to the remaining tenets of the evolutionary psychology paradigm—to its claims about mind, methodology, metatheory and policy.

The juncture where evolutionary psychology as a paradigm begins to break down is precisely the juncture that perhaps tempted some evolutionary psychologists to board the rapid train of thought towards evolutionary psychology as being conceived of a revolutionary paradigm in the first place. A strong consensus has emerged that *a priori* reasoning about selection simply cannot demonstrate that the mind is more likely to be massively modular: selection is consistent with an assortment of different possible cognitive architectures. Considerations of empirical adaptationism simply do not speak against or destabilise domain-general mechanisms or intelligence. This alone is sufficient to end evolutionary psychology's call to be recognised as the metatheory for the behavioural sciences.

This doesn't, of course, rule out the possibility of psychological adaptations. And the failure to establish *a priori* the likelihood of massive modularity does not speak against the legitimacy of evolutionary psychologists hypothesising about adaptive problems and adaptive solutions. Nevertheless, sceptics identify multiple problems and limitations with adaptationist methodology in psychology. There appears to be a degree of indeterminacy with the method, to the extent that it seems to be unconstrained. Furthermore, the explanations it produces are wafer thin, lacking the deep details expected of reasonable adaptationist explanations. And it seems to be a complete mystery how evolutionary psychology could be heuristic.

These considerations—and not purported tensions between evolutionary psychology and various theoretical developments—are catastrophic to evolutionary psychology as a paradigm: they expose the hollowness of talk about it unifying not only psychology, but also the behavioural sciences, and evolutionary psychologists' efforts to draw policy implications and to popularise their work starts to look potentially dangerous. Indeed, it is precisely at the juncture of policy that the language against evolutionary psychology becomes the ugliest and the debate the most unpleasant. Accordingly, in 2.6, 2.7, 2.8 and 2.9 we shall examine the failure to infer massive modularity from empirical adaptationism, key methodological objections, the vacuousness of metatheory talk, and potential dangers concerning policy and popularisation respectively.

As you can expect, all this has led to a powerful impression that evolutionary psychology can be written off wholesale. Ultimately, I don't believe the hard sceptics are entitled to that judgement because evolutionary psychology methodology is more sophisticated and useful than they appreciate. But they nevertheless pose legitimate questions and they need to be understood before answering. And it will take a little while to answer those methodological objections because they require us to try to get clearer not only on the details of adaptationist methodology in psychology, but also the fundamental rationale for adaptationist hypothesising in psychology. We shall explicate and discuss that matter in the next chapter.

2.1 Levels of Selection

The levels of selection question—the question of whether natural selection operates on individuals, genes, groups or some other unit—has been, and remains, a lively topic of debate in the literature. Various positions on this question have been championed in the literature. Individual selection was championed by Darwin (1859); kin selection, a term coined by Maynard Smith (1964), was established by Hamilton (1964) and Williams (1966), strengthened by Price (1970) and popularised by Dawkins (1976); classic group selection was proposed by Wynne-Edwards (1962), though this thinking can also be found in Darwin's works; and neo-group selection was established by Sober and Wilson (1999).

The level of selection question has typically been discussed with reference to the problem of altruism. Biological altruism is defined as behaviour that benefits someone else at a cost to oneself (where costs and benefits are calculated in terms of reproductive fitness). Biological altruism is widespread in nature. Classic examples include ant workers foregoing personal reproduction to help the queen reproduce; lionesses nursing cubs that are not their own; ground squirrels warning each other of the approach of enemies (Trivers, 1985); when attacked by wolves, musk oxen 'wagontrail', forming a circle while the females and young shelter in the circle's interior (Sober, 2000); and bees disembowel themselves when they sting intruders to the nest.

Darwin recognised that altruism poses a problem for the theory of evolution by natural selection. He was troubled by the phenomenon of sterile workers in insect colonies, which devote their lives to helping a queen reproduce at the expense of having offspring themselves. If evolution by natural selection is understood to favour individuals who increase their own reproductive success relative to other members of the same species, then it's something of a puzzle how altruistic traits, traits that aid the reproductive success of other members of the same species at the reproductive cost of the altruistic individual concerned, could have been preserved by natural selection. Indeed, on this basis, it would seem that altruistic traits would be selected against.

Darwin tentatively accounted for this problem by suggesting that selection can also act at the level of the group. Even though self-sacrifice reduces the reproductive fitness of the individual concerned it can nevertheless be beneficial for the reproductive fitness of the group of which the individual is a member. Darwin hypothesised that tribes which contain members that sacrifice themselves for the good of the group will survive and reproduce at a greater rate than tribes that don't contain such members. Competition between altruistic and non-altruistic groups could, therefore, have led to the selection and distribution of altruistic traits.

In the decades immediately after Darwin, understanding natural selection as operating at the level of the group became the orthodoxy. Altruism didn't command notable scientific interest as it was not seen as an anomaly. Those who believe natural selection operates at the group level argue that natural selection favours traits that benefit the group as a whole. Thus, evolution works for the 'good of the group'. Hence, altruism was only to be expected.

Group selection reached its classic exposition in Wynne-Edwards (1962). Wynne-Edwards proposed that many species have mechanisms ensuring that groups do not exhaust their resource base. These mechanisms evolved through group selection. Populations without such mechanisms are liable to erode their resource base and thus risk extinction. Infanticide in langur monkeys is an example of a trait claimed to have evolved for the good of the group. Dominant male langur monkeys who take over harems of females frequently kill all infants under six months of age, infants still suckling. This has been claimed to be population control for the good of the group, as it dramatically limits and regulates population growth, limiting the risk of overpopulation.

In the 1960s, however, the 'good of the group' consensus in evolutionary biology was destabilised and displaced. Maynard Smith (1964) and Williams (1966) argued that group selection is, at best, a weak evolutionary force—group selection will only generate significant effects under very special conditions, conditions that rarely obtain outside experimental settings. Classic group selection is vulnerable to the free rider theorem, also known as subversion from within, which demonstrates that altruistic groups are highly vulnerable to subversion from within. Suppose organisms within a group are in competition with one another. Now suppose there are two types of organism within the group: those who act for the good of the group, and those who do not. Selfish organisms receive the benefits that obtain from altruistic behaviours but bear none of the costs. Selfishness is fitter, and so will eventually drive altruism in the group to extinction. Even if the group has only altruistic organisms, a single selfish mutant is sufficient to drive altruism in the group to extinction.

If classic group selection is unlikely to account for how altruism evolved, what could? Kin selection and reciprocal altruism were formulated and championed instead.

Kin selection is selection due to interactions between kin, the process whereby traits are selected for because of their beneficial effects on relatives' fitness. In a nutshell, altruism can evolve if the cost to the altruist is offset by sufficient benefit to sufficiently related recipients (Hamilton, 1964). This idea can be expressed in a simple equation. In its basic form, Hamilton's rule is $Br>C$, where B represents the benefit to a gene for altruism in the recipient; r is the degree of genetic relatedness; and C is the cost to an identical copy of the gene in the altruist.

West *et al.* (2007) helpfully distinguish between the narrow use of kin selection and the broader use of kin selection. 'The narrower use of kin selection works upon interactions between individuals who are genetically related due to common ancestry—i.e. indirect benefits due to limited dispersal or kin discrimination' (*ibid.*: 417). Relatedness, of course, can occur without recent common ancestry. A greenbeard gene can identify, and assist, copies of itself in other individuals. This opens up a broader definition of kin selection: 'The broader use of kin selection works upon interactions between individuals who share the gene of interest, regardless of whether this is due to coancestry or some other mechanism—i.e. also includes greenbeard effects' (*ibid.*: 417). Hence, the difference between the narrow and broad uses of kin selection is 'whether kinship and relatedness are defined on the basis of average genetic similarity over most of the genome (narrow definition), or at the particular locus of the behaviour being examined (broad definition)' (*ibid.*: 417).

Organisms should behave in ways that maximise their average lifetime inclusive fitness, a measure taking into account not merely the number of direct offspring (personal fitness), but also the offspring of relatives. More precisely, 'An organism's inclusive fitness is defined as its personal fitness, plus the sum of its weighted effects on the fitness of every other organism in the population, the weights determined by the coefficient of relationship r' (Okasha, 2008a).

What about the evolution of altruism towards non-kin? This was to be explained by Trivers's (1971) theory of reciprocal altruism. Trivers' basic idea was straightforward: it may pay an organism to help another, if there is an expectation of the favour being returned in the future ('If you scratch my back, I'll scratch your back'). More precisely, an organism behaving altruistically towards a non-related member of the same species will increase its reproductive success if there is a probability of an altruistic act being returned to it by the recipient organism sometime in the future. The reproductive cost to the organism of behaving altruistically is compensated by the probability of the return in reproductive benefit, thereby allowing the behaviour to evolve by natural selection. For reciprocal altruism to work, it's necessary that individuals interact with each more than once, and have the ability to recognize other individuals with whom they have interacted in the past.

During the 1970s and 1980s, the concept of group adaptation was thrown 'into the outer darkness' (Sober, 2000: 107). Kin selection and reciprocal altruism became the new orthodoxy in evolutionary biology, and the pioneering evolutionary psychology practitioners in the 1980s and early 1990s inherited this orthodoxy. However, during the mid-to-late 1990s, group selection was reformulated, into what is now known as new or neo-group selection. Sober and Wilson (1999) demonstrated how group selection can explain the evolution of altruism, that there are circumstances relating to population structure where altruism can evolve by group selection—which, crucially, demonstrates how subversion from within is not insurmountable for group selection. Sober and Wilson argued that the fact that in mixed groups the fitness of altruists is lower than the fitness of selfish individuals doesn't mean that overall the number of altruists cannot go up. Put another way, at the individual level, the selfish trait will be favoured. However this does not mean that, at the group level, selfish groups will be favoured.

For group selection to work, several conditions need to obtain. One condition is that groups must differ in the number of altruistic members they contain. Another condition is that groups must not be stable, but must continually form and reform (and at a sufficiently fast rate). Classic group selection postulated well-defined groups. In contrast, new or neo-group selection has arbitrarily defined groups. From this perspective, despite being selected against under individual selection, altruism can, indeed, evolve because altruists tend to associate with one another. We can compare groups of altruists against other groups, and observe that the altruistic groups are the fittest, and that therefore altruism can spread. If groups break apart to form new groups frequently enough, then selfishness will not have the opportunity to spread. Altruism will prevail.

Although group selection remains contentious in more orthodox circles, it's generally agreed that Sober and Wilson's arguments have reinvigorated, and renewed interest in, group selection (Borrello, 2005; Gangestad and Simpson, 2007; Okasha, 2008b). However, despite this reformulation, reassessment and recognition of group selection, evolutionary psychologists still tend to utilise kin selection and other classic evolutionary theories, not neo-group selection. This has been criticised. Lloyd and Feldman (2002) and Laland and Brown (2002) note that evolutionary theory has 'evolved', that evolutionary theorists and geneticists now use multi-level selection models, and they accordingly criticise evolutionary psychologists for becoming detached from such recent developments in evolutionary thinking. Gintis (2009: 244) claims that 'Evolutionary psychology ... has incorporated the kin selection/reciprocal altruism perspective into a broadside critique of the role of culture in society'. Gintis then dismisses evolutionary psychology because of its adherence to the kin selection/reciprocal altruism framework.

It's a mistake to suppose that the levels of selection debate problematises evolutionary psychology. First, one could pursue evolutionary psychology along neo-group selection lines without difficulty. Second, kin selection and neo-group selection are actually different ways of looking at the same thing, and there are good reasons to prefer kin selection over neo-group selection.

First, suppose one argues that neo-group selection is 'correct', that kin selection is 'incorrect' and that evolutionary psychologists should adopt neo-group selection. Even if evolutionary psychologists were moved by such an argument, which seems unlikely for reasons to be articulated in a moment, they could quite happily adopt neo-group selection without revising or overhauling evolutionary psychology's theoretical and methodological foundations. Whether selection occurs at one level or another level or multiple levels, adaptations will always be located at the individual level (Davies *et al.*, 2012). Whether one adopts kin selection or neo-group selection, adaptationist questions and hypotheses can still be formulated about psychological traits within individuals. In principle, evolutionary psychologists could investigate, and try to find evidence for, group selection on the design of purported adaptations, as traits that evolved by group selection should have features designed to function at that level. Indeed, Kurzban and Aktipis (2007: 230) invite us to consider the possibility of detecting 'the footprints' of group selection in the mind: 'it is possible that certain mechanisms that generated behaviors leading to greater assortment were selected because they increase the strength of group-level selection. In other words, traits that caused cooperative individuals to group together probably increased the likelihood of selection at the group level.'

Second, it's not actually the case that we need to declare kin selection correct and neo-group selection incorrect or vice versa. Recently, a consensus has emerged that kin selection and neo-group selection are formally or mathematically equivalent (Lehmann *et al.*, 2007; Okasha 2008a, 2008b; Kohn 2008; Marshall 2011). They are 'intertranslatable' (Wilson and Wilson, 2007). According to West *et al.* (2008: 378),

> kin selection and (new) group selection are mathematically equivalent ways of looking at the same thing. We cannot emphasize strongly enough that it is not the case that one is correct and the other wrong, nor that group selection predicts things that cannot also be predicted with kin selection theory.

So what we have here is a case of underdetermination: both frameworks, mathematically equivalent, can explain the same range of phenomena. Claiming that organisms should behave in ways that maximise their average lifetime inclusive fitness is consistent with claiming there are multiple levels of selection.

Even if both frameworks are formally equivalent, evolutionary psychologists can have good reasons for adopting kin selection over neo-group selection. For example, one could argue that kin selection is more parsimonious than neo-group selection. Indeed, one could argue that kin selection is the more useful framework (perhaps precisely because it's more parsimonious). West *et al.* (2008: 381) make the important point that the two frameworks being formally equivalent doesn't mean they're equally useful:

> At one level, kin selection and group selection are just different ways of doing the maths or conceptualizing the evolutionary process. However, from a practical point of view, it could not be clearer that the kin selection approach is the more broadly applicable tool that we can use to understand the natural world. This is because kin selection methodologies are usually easier to use, allow the construction of models that can be better linked to specific biological examples, lend themselves to empirical testing and allow the construction of a general conceptual overview.

So according to this characterisation, which, admittedly, champions of neo-group selection might seek to resist, evolutionary biologists concerned with modelling animal behaviour find kin selection models far more useful than group level models. Indeed, West *et al.* (2007: 424) claim that important discoveries made using kin selection would not have been as easily made using neo-group selection:

> in some of the most successful areas of social evolution, such as split sex ratios in social insects or extensions of Hamilton's (1967) basic local mate competition theory, predictions arise elegantly from kin-selection models, whereas the corresponding group selection models

would be either unfeasible or so complex that they have not been developed.

2.2 Environment of Evolutionary Adaptedness

In considering the adaptive problems faced by our ancestors, evolutionary psychologists consider our environment of evolutionary adaptedness (EEA). The EEA refers those aspects of our ancestral environment that were relevant to the evolution of our adaptations (Hagen, 2005). More precisely,

> The 'environment of evolutionary adaptedness' (EEA) is not a place or a habitat, or even a time period. Rather, it is a statistical composite of the adaptation-relevant properties of the ancestral environments encountered by members of ancestral populations, weighted by their frequency and fitness-consequences (Tooby and Cosmides, 1990: 386–7).

So, strictly speaking, the EEA is neither a time nor a place. In practice, a few evolutionary psychologists occasionally refer to our environment of EEA as the Pleistocene, a period of time approximately 1.8 million to 10,000 years ago (e.g. Hagen, 2004; Tooby and Cosmides, 2005). As Tooby and Cosmides (1990: 388) explain,

> for most ordinary analytic purposes, the EEA for a species (i.e., for its collection of adaptations) can be taken to refer to the statistically weighted composite of environmental properties of the most recent segment of a species' evolution that encompasses the period during which its modern collection of adaptations assumed their present form. We have used the word 'Pleistocene' in this sense to refer to the human EEA, because its time depth was appropriate for virtually all adaptations of anatomically modern humans.

The thought here is that our ancestors evolved during the Pleistocene into the modern human; thus, psychological characteristics that are distinctly human should have evolved and solidified during this period, hence the references to the Pleistocene occasionally found in some parts of the literature.

The EEA concept is part-and-parcel of the concept of adaptation. One cannot describe an adaptation without describing an ancestral environment to which the adaptation is adapted. Hence, each adaptation has its own EEA, a statistical aggregate of selection pressures responsible for the emergence the adaptation (Buss *et al.*, 1998). For example, the EEA for the human eye will differ from the EEA for the human language faculty (Tooby and Cosmides, 1990).

Selection pressures need to be sufficiently stable to give rise to specific adaptive problems. Crucially, the physical environment need not remain constant. Sometimes, one reads objections to evolutionary psychology that boil down to 'the EEA had variable weather'. But that misses the point. The adaptive problem of finding a mate, for example, was a consistent, recurring selection pressure, regardless of environmental variables (and remains so).

The EEA concept encourages researchers to recognise that a selected trait is adaptive to its EEA, not necessarily to today's environment. Post-Pleistocene agriculture- and industry-based societies are very different from Pleistocene hunter–gatherer societies. There has been rapid technological and cultural change. Certain features of our environment are evolutionarily novel. Think modern contraceptive interventions to limit fertility. Think speedy international travel. Think eating habits. This mismatch means a selected trait might not have current utility. Indeed, it might be deleterious, either in the technical sense of reducing genetic fitness or in the everyday sense of reducing health and well-being.

There are clear and striking examples of mismatches between adaptations and today's environment. Perhaps the best known example is our strong preference for fatty and sugary foods. Despite knowing the health risks associated with eating too many fatty and sugary foods, many continue to do so and find it difficult to stop. It is commonplace to observe that such foods are 'addictive'.

Why do we possess such a strong desire for fatty and sugary foods? The well-known adaptationist explanation runs as follows. As hunter–gatherers, our ancestors' food supplies were likely to have been volatile—certainly a far cry from today's all-too-predictable three meals a day and multiple snacks. Food shortages were likely to have been regular occurrences. In such a situation, a strong preference for fats and sugars would have been adaptive, as consuming high-energy fats and sugars would have supplied the energy needed to survive such restricted energy periods. Indeed, given the relative scarcity of fats and sugars in ancestral environments, gorging on such high-energy foods when they became available would have been critical. Hence, our preference for fatty and sugary foods can be understood as an adaptation to recurrent periods of calorie/energy restriction.

These fat and sugar preferences remain with us today. But instead of finding ourselves in a world where fats and sugars are relatively scarce, we now find ourselves in a world where foods rich in fats and sugars are plentiful. Indeed, our evolved preferences explain why such foods are so widespread—such foods 'sell well precisely because they correspond to, and exploit, evolved desires for these substances' (Buss, 2008: 65). However, the strong preferences for fat and sugar that served our ancestors so well are shockingly dysfunctional in today's world, leading to obesity and a cascade of health problems such heart disease and diabetes.

Fear of snakes is another vivid example of mismatch between aspects of our psychology and our modern environment. Snakes strike terror. It is commonplace that, when in a natural environment, unexpected, small, sudden movements in a bush or grass will captivate attention and refocus our conscious awareness. We are seized by fear. We step back. We fear the possibility of a snake in the background.

But we don't have to find ourselves in a natural setting in order to experience the fear of snakes. The thought of them elicits fear. The look of them. The presence of them. Their movements, their slithering, their scales, their colours, as if they're designed to elicit dread in us—or rather, as if our psychology is designed to respond to such visual cues.

Yet, snakes on a plane aside, snakes simply do not represent a credible threat to people in urban environments. Their existence simply doesn't warrant the kind of dedicated psychological system that seems to be dedicated to screening for them. Cars and guns, however, do represent genuine and credible threats. If one were designing motivational mechanisms for humans in today's environment, one would certainly dial down or switch off the precious cognitive space devoted to a fear of snakes in favour of installing new mechanisms for car and gun fears.

It's irrational to be hypersensitive to cues of statistically remote threats but not to statistically closer threats. We seem to have a psychological solution to a problem that no longer obtains and no psychological solution to a problem we do face. But of course snakes were ubiquitous in an ancient environment, in our evolutionary environment. They were once a genuine threat, to an extent that having a functionally specialised cognitive system dedicated to that threat would have been advantageous.

Look at alarm or fear from an adaptationist perspective. When faced with threats, fear is activated, redirecting and recalibrating various systems—perception, the ranking of goals, behavioural decision rules—in order to protect an organism from a perceived threat. Cosmides and Tooby (2000) nicely trace out the steps involved. Suppose you are walking alone at night; you hear a noise behind you. This noise acts as a cue to the possibility of a threat; fear is triggered, seizing and recalibrating various systems: the threshold for signal detection is lowered; there are shifts in perception and attention (greater sensitivity to sounds that could indicate the presence of a threat, becoming aware of sounds normally not picked up on); there is a change in the weighting of goals (safety rockets up the priority list, food and mate-finding become lower priorities, and hence one no longer feels hungry, thirsty or sexually aroused); memory systems are activated (safe places and the location of allies are recalled); physiology changes take place (adrenalin spike, blood rush, heart rate increase); and behavioural decision rules are activated (which are sensitive to the nature of the perceived threat—to fight, to run, to cry or to be still).

The crucial point concerns the kind of cues or stimuli that trigger fear and alarm—this is where evolutionary psychology can yield important insights. As we know, in virtue of its emphasis on specificity of adaptations, evolutionary psychology suggests that the existence of a general 'threat detection' fear mechanism is unlikely. There were no general threats in the EEA; threats were concrete, particular, in flesh, scales and blood, in firestorms and thunderstorms; consequently, cues that the fear mechanism process will relate to these threats. So evolutionary psychology can yield the following expectation: the closer a threat is framed in terms of, or approximates, a threat faced by our ancestors in the EEA, the more likely it will be to trigger alarm.

What threats did our ancestors face? One of the major threats, perhaps the chief threat, came from hostile fellow humans, whether members of the same tribe or a different one. Consequently, we can expect cognitive systems to have evolved that are very sensitive to threatening cues, such as facial expressions. Indeed, various studies have demonstrated that angry faces are detected more quickly than non-angry faces (e.g. Hansen and Hansen, 1988). Furthermore, there is an interesting sex difference here, one discovered by evolutionary psychologists—although men and women identify facial expressions of anger more quickly than other emotions, men register expressions of anger on male (but not female) faces faster than women do (Williams and Mattingley, 2006). On the basis that violence is mostly male-on-male, and probably has always been so, the adaptationist perspective predicted, and confirmed, this sex difference. Indeed, this nicely exemplifies how selective forces often shape very specific adaptations because very specific needs were present in past environments.

Another major source of danger came from predators, as well as dangerous animals. Our psychologies should be sensitive to predator threats, even if those threats cease to obtain in modern suburban environments. This is indeed the case: it's far easier to not only elicit, but also to further condition, a fear of snakes in people than it is to condition a fear of cars, which represent a greater threat to people today than snakes do. Finally, our ancestors faced ecological threats and disasters—floods, storms, landslides, wild fires, heat waves, droughts and famines. Indeed, there was an ice age around 13,000 years ago.

So the way our alarm processes are calibrated is rather like a smoke detector that's calibrated to detect only a certain type or range of smoke: it might work well most of the time when we're asleep, but should a type or range of smoke that's not calibrated emerge, the alarm won't be triggered.

Hence, we have dedicated mechanisms to detect and respond to threats faced in our ancestral environment—not just to snakes but the wide spectrum of threats recurrently encountered. 'The other common fears are of heights, storms, large carnivores, darkness, blood, strangers, confinement, deep water, social security, and leaving home alone. The common thread is obvious. These are the situations that put our evolutionary ancestors in danger' (Pinker, 1997b: 386).

In stark contrast, we do not have dedicated mechanisms to elicit evolutionary novel threats like guns and cars. And perhaps we also don't have a dedicated mechanism to face a key threat we might now be facing, a possible adaptive problem that has no parallel in our evolutionary past, an evolutionarily novel adaptive problem that possibly threatens to change our environment dramatically.

Regardless of the controversies of whether climate change is happening or not and, if so, how to model it, the fact remains that many people do believe it's happening and yet those same people are simply not alarmed at the emotional level. People can acknowledge and register climate change as a threat, but such a threat rarely, if ever, elicits alarm.

The adaptationist perspective explicated above offers a key insight: climate change does not present the kind of cues or stimuli that the mechanisms dedicated to responding to threats were evolved to respond to. Most obviously, climate change is not a hostile face or a predator. However, it is an ecological threat, so perhaps we should expect climate change to trigger the alarm system? The problem here is that climate change, if correctly modelled, is taking place at a rate below the threshold needed to trigger alarm. Such a slow-paced but looming threat of ruin would be evolutionarily novel.

Compare the lack of alarm over climate change with the alarm trigged by the discovery of the rapid disappearance of the ozone layer. The story usually goes as follows. In 1983, Joe Farman, using a 25-year-old machine in a research station in Antarctic, discovered that half the ozone layer over Antarctica appeared to have vanished. Together with Brian Gardiner and Jon Shanklin, Farman announced his discovery in a paper in *Nature* in May 1985. Initially, there was disbelief: not surprising, as a NASA satellite had been measuring ozone levels for several years and didn't detect significant depletion. Nevertheless, NASA reviewed its satellite findings. Incredibly, the satellite had detected the massive depletion, but its data quality control algorithms had automatically rejected the data as being wildly improbable. Once the data were re-run without the data quality control algorithms, the massive depletion was verified.

The sudden and dramatic discovery of 50 per cent depletion of the ozone over Antarctica shocked the scientific community. It caused international alarm—and action. The Montreal Protocol, a treaty phasing out the chlorofluorocarbons (CFCs) responsible for the ozone depletion, was drafted only two years after the *Nature* paper, and came into force in 1989. The ozone layer is now slowly repairing. The threat associated with ozone depletion—the threat of lethal ultraviolet solar rays—was dealt with quickly. Not one United Nations member state has failed to sign up to protocols, and the global production and consumption of ozone-depleting substances has dramatically deceased to the point of safety. No ifs, no buts, no passing the buck, no appeals to tragedy of the commons—just speedy, decisive, uniform action on an unprecedented scale.

The difference between the ozone hole over Antarctica and climate change is that the former was a massive change at a rapid rate, while the latter is taking place slowly. Because of the speed, it triggered alarm in a way climate change so far hasn't and might not do until it's too late. Of course, the ozone layer depletion was also evolutionarily novel, but one fast enough to trigger the alarm circuitry.

We often hear that the tragedy of climate change is that it's taking place too quickly. If the adaptationist perspective here is right, perhaps the real tragedy of climate change is actually that it's taking place too slowly. If it were more sudden, if (for example) the Arctic became ice-free abruptly, rather than gradually over a stretch of time, then this would likely trigger alarm, and orchestrate speedy international action to counteract or mitigate the threat. A sudden and dramatic ecological change would do more to animate climate change countermeasures than all the summits, education programmes and government initiatives have achieved in 20 years, or are ever likely to achieve alone, without our alarm processes being trigged.

So people acknowledge and register the statistics of climate change at an intellectual level but it's clearly difficult to respond to it at the emotional level—and it's reasonable to say that emotions are what really drive us. This, I think, goes some way to illuminating the puzzlingly situation of belief in an ecological threat like climate change failing to trigger alarm, especially puzzling when ecological threats in the past like ozone layer depletion have triggered alarm. I should stress that I'm not aware of any evolutionary psychologist who has examined this particular issue but it seems quite clear that, from an adaptationist perspective, climate change or belief in climate change doesn't represent the kind of input that will trigger our adapted alarm systems in virtue of climate change's relatively slow pace of change.

So we've established that some of our adaptations appear to be strikingly mismatched with today's environment. Crucially, however, one must distinguish the possibility of mismatch from the ubiquity of mismatch. One can endorse the possibility that psychological adaptations can be mismatched with today's environment without endorsing the claim that the majority of psychological adaptations are mismatched with today's environment. Symons (2005: 226) makes the crucial point that the EEAs of many of our adaptations are likely to be continuous with the present:

> The EEAs of the vast majority of human adaptations still exist today and usually are too obvious to merit explicit mention. For example, a neurophysiologist describing the function of a certain component of the human visual system probably will simply assume that his or her colleagues know: (1) a great deal about the nature of electromagnetic radiation and (2) that the (natural) light falling on human retinas today is essentially identical to the light that fell on our ancestors' retinas during the evolution of our visual system.

Hence why humans so successfully live and flourish in a variety of social environments. Most of our adaptations, physiological and psychological, work splendidly precisely because their EEAs are continuous with the present. Hence also why cases of maladaptivity are so striking—they're striking precisely for their relative rarity, a Darwinian 'man bites dog'.

Yet this distinction is rarely made. People confuse possibility with ubiquity. Both popular accounts and detractors of evolutionary psychology tend to think evolutionary psychology as holding that humans are cavemen bumbling about in a strange and weird world. For example, *Psychology Today*, perhaps the most influential psychology website on the internet, features website subsections called 'Ape Girl', 'Caveman Logic' and 'Caveman Politics', subsections written by professional psychologists on evolutionary psychology topics. The blurb for 'Caveman Logic' is to examine how 'our *primitive* minds are mismatched to the modern world' (my emphasis). We also read, 'Human instincts were designed for hunting and gathering on the savannahs of Africa 10,000 years ago. Our present world is incompatible with these instincts because of radical increases in population densities, technological inventions, and pollution' (Barrett, 2010: 3; not to be confused with H. C. Barrett).

Little surprise, then, that sceptics end up absorbing this picture and then seek to challenge it. Buller (2005: 112) reassures us:

> There is no reason to think that contemporary humans are, like Fred and Wilma Flintstone, just Pleistocene hunter–gatherers struggling to survive and reproduce in evolutionarily novel suburban habitats.

Dupré also finds it 'quite surprising' that we are 'systematically maladapted' and that there is 'no good reason' to accept the idea (2012: 246). Agreed. But I doubt one can find a major evolutionary psychologist who would disagree.

Another issue related to the EEA is the possibility of post-Pleistocene selection for psychological adaptations. Referring to the 10,000 years since the end of the Pleistocene period, Barkow *et al.* claim that 'it is unlikely that new complex designs ... could evolve in so few generations' (1992: 5). Our psychological adaptations are complex traits, and the construction of complex adaptations perhaps requires tens or even hundreds of thousands of years of cumulative selection. Our species spent over 99% of its evolutionary history as hunter–gatherers. Agriculture, which changed everything, has only been around for 10,000 years; industrialisation for only for 200. The few thousand years since the advent of agriculture is a tiny stretch in evolutionary time. 'Therefore, it is improbable that our species evolved complex adaptations even to agriculture, let alone to postindustrial society' (*ibid.*).

Some have challenged this claim. Sociobiology, the forerunner of evolutionary psychology, did not make this assumption. In fact, Wilson (1975: 569) warned precisely against it:

> There is no reason to believe that during this final sprint there has been a cessation in the evolution of either mental capacity or the predilection toward special social behaviors. The theory of population genetics and experiments on other organisms show that substantial changes can occur in the span of less than 100 generations, which for man reaches back only to the time of the Roman empire [...] it would be false to assume that modern civilizations have been built entirely on capital accumulated during the long haul of the Pleistocene.

Indeed, in recent years evidence has emerged that there has been post-Pleistocene selection. For example, Mekel-Bobrov *et al.* (2005) found that a variant of the gene *ASPM*, which is a regulator of brain size, arose just 5,800 years ago and is now carried by about a quarter of the world's population. According to the researchers, these findings 'suggest that the human brain is still undergoing rapid adaptive evolution' (*ibid.*: 1702). Another example is provided by Lewens (2007), who notes that the increase in the use of dairy products resulting from the domestication of cattle has resulted in a significant increase in lactose tolerance. Lewens (2007: 154) draws a lesson from this:

> Of course, the invention of dairy farming is only one of the many changes wrought on our environment since the Pleistocene. Many of us now live in cities, we no longer hunt as a matter of necessity, medical technology has improved, and there is no reason to rule out considerable modification of our cognitive adaptations in response to these altered environments, too.

Laland (2007) draws an even stronger conclusion. Laland argues that, back in the early 1990s, it was tenable for Tooby and Cosmides to hold that selection is typically slow, and hence it was only reasonable to suppose that little selection has occurred since the end of the Pleistocene. However, we now know that rates of genetic evolution can be much faster. On the strength of these developments, Laland claims that 'the theoretical framework underpinning narrow evolutionary psychology is untenable' (2007: 9).

I believe too much has been made of this issue. Perhaps it's possible for post-Pleistocene psychological adaptations to have arisen, but to place so heavy an emphasis on one relatively small category of possible psychological adaptations, a category of adaptation that so far is empty of actual examples, is to get its true significance out of perspective. To bring out this thought, let's set out a number of responses.

First, evolutionary psychology is not committed to the claim that there has been no post-Pleistocene evolution. That would be an unnecessarily strong claim to make. Rather, evolutionary psychologists seem to make a much more modest claim: that there hasn't been sufficient time for significant changes or additions to our psychological adaptations.

Kurzban (2011) emphasises that there is a crucial difference between rapid genetic evolution and the evolution of complex adaptations. Evolutionary psychologists are concerned with the later, not the former. Furthermore, evidence of recent genetic evolution does not constitute evidence for the possibility, let alone the actuality, of post-Pleistocene complex psychological adaptations. Indeed, the reasoning that 10,000 years since the advent of agriculture is insufficient time to select for new complex psychological adaptations remains untouched by evidence of recent, rapid genetic change. An evolutionary psychologist can happily acknowledge rapid genetic evolution while comfortably denying the existence of post-Pleistocene complex adaptations.

Second, unsurprisingly given what we've just said, no evidence of significant changes to our psychological architecture since the end of the Pleistocene has been forthcoming. The evidence cited relates to simple physiological traits, not complex psychological adaptations. Lewens claimed there is no reason to rule out considerable modification of our cognitive adaptations in response to altered environments but he declined to identify any possible candidates.

Third, let's assume that a few thousand years is sufficient time for selection to engineer new complex psychological adaptations. However, this alone is insufficient to generate new psychological adaptations. Consider the type of diet consumed in English-speaking countries. If average Anglo-Saxon diets continue to be nutritionally impoverished, continue to be laced with fructose, if obesity levels continue to escalate in America and the UK, this might lead to a selective advantage for some kind of fructose intolerance. In lands of obesity, a mild form of fructose intolerance might offer a selective advantage. But if Anglo-Saxon diets change, if they became more Mediterranean-like, more balanced, wholesome and nutritious, then the pressure for fructose intolerance would evaporate.

The point is that for the new environments we have faced in the last 10,000 years to create new adaptive problems, these environments need to be stable enough to generate stable adaptive problems. But, arguably, there's been very little stability in societal arrangements in the last few decades, let alone centuries, let alone thousands of years. Mediaeval

England is a different world to Victorian England. The England of my childhood is not the England of my twenties, and it will not be the England of my forties, or sixties, nor, if I reach it, my eighties and beyond. There is massive flux, massive re-orderings and configurations and developments of society, of economics, of technology, of social norms, culture and ideas. In short, there has perhaps been too much instability in social structures since the Pleistocene to create novel adaptive problems requiring novel solutions. Perhaps societies will eventually settle into recurring novel patterns, perhaps not. But there's little sign yet of ever-accelerating cultural change slowing down.

Fourth, let's be really generous, and grant both the possibility of post-Pleistocene selection for psychological adaptations and the likelihood or actuality of post-Pleistocene psychological adaptations. Does this discredit evolutionary psychology? Is this game over for evolutionary psychology? Absolutely not. Such considerations would constitute a new research programme, one dedicated to hunting for post-Pleistocene psychological adaptations, one that can cheerily enough run alongside evolutionary psychology. Even if we have new psychological adaptations, this wouldn't undermine the claim that we have, or might have, psychological adaptations dedicated to solving recurring adaptive problems in the ancestral past, problems with ancient pedigrees.

So no matter which way one looks at it, the issue of recent selection doesn't problematise evolutionary psychology. Certainly, it doesn't make evolutionary psychology's framework 'untenable'.

2.3 Behavioural Flexibility

Although evolutionary psychologists seek to understand human behaviour, they do not focus on selection as acting directly on specific behaviours. Rather, they focus on selection acting on information-processing psychological mechanisms. Of course, in practice, evolutionary psychologists often speak of behaviours as being adaptations—such as jealousy as an adaptation. Nevertheless, this is merely shorthand, sparing readers more cumbersome phrases like 'the mechanism underlying or generating the behaviour is an adaptation'.

Focusing on selection and psychological mechanisms, instead of selection and behaviour, is what crucially distinguishes evolutionary psychology from sociobiology: 'this new field focused on psychology—on characterizing the adaptations comprising the psychological architecture—whereas sociobiology had not. Sociobiology had focused mostly on selectionist theories, with no consideration of the computational level and little interest in mapping psychological mechanisms' (Tooby and Cosmides, 2005: 16, n. 3).

If one applies adaptationist theorising straight to the level of manifest behaviour, one risks becoming blind to behavioural variation. Indeed, it might tempt one to average out variation. Referring to previous evolutionary research that applied direct to the level of manifest behaviour, Cosmides and Tooby (1987: 279) note that 'often the researcher would take the observed variation, average it, and typify the species or group by that average'. A possible recent example of this is Okasha (2007). Okasha attempted to provide an evolutionary explanation for the pervasiveness of risk-aversion in decision making. Schulz, however, rightly points out that the focus on risk *aversion* alone fails to do justice to the diversity and complexity of risk attitudes:

> The real issue concerning human attitudes towards risk is not that people are risk-averse, but that they are risk-averse and risk-loving. More specifically, it is the fact that many people are both risk-loving and risk-averse at the same time. Explaining this non-monotonicity in people's attitudes towards risk is the real heart of the problem—making sense only of their risk-aversion merely covers half of the ground to be traversed' (2008: 160).

And surely that is right. In some conditions, I might be risk friendly, and another person risk averse. In alternative conditions, I might be the one risk friendly, and the other person risk averse. An evolutionary account should account for this diversity, not average it out.

Early sceptics of applying evolutionary perspectives to behaviour highlighted and emphasised the extraordinary complexity and diversity of human behaviour, and then contrasted this with what they judged to be naive, simplistic adaptationist accounts. Even sympathisers of evolutionary accounts of behaviour shared this concern. For example, surveying the contemporary sociobiological work of his day, Dawkins (1976: 191) said they 'are plausible as far as they go, but I find that they do not begin to square up to the formidable challenge of explaining culture, cultural evolution, and the immense differences between human cultures around the world'. Hence, Dawkins instead opted for the cultural evolution route for explaining social behaviour, in particular formulating his own theory of memetics.

Focusing our attention on selection acting on complex psychological traits, instead of manifest behaviour itself, enables the possibility of properly accounting for some of the undeniable complexity and diversity of human behaviour. Again, we can appeal to the symmetry principle. Just as various physiological adaptations function across a very wide variety of environmental settings—the immune system responds to a variety of pathogens, we can digest a variety of foods—so too should our psychological adaptations function across a wide variety of ecological and cultural settings (Barkow, 2006).

And this is precisely what happens with non-humans. Krebs (2005: 752) notes that members of species ranging from crickets and crayfish to chimpanzees have been found to adopt conditional strategies such as, 'If your opponent seems more powerful than you, defer to him or her; if your opponent seems less powerful than you, intimidate him or her'. Daly and Wilson (2001: 4) also inform us that

> Experimental studies of nonhuman animal foraging decisions have established the ecological validity of such a risk preference model. Rather than simply maximizing the expected (mean) return in some desired commodity, such as food, animals should be—and demonstrably are—sensitive to variance as well (Real & Caraco, 1986). For example, seed-eating birds are generally risk averse, preferring a low variance foraging situation over one with a similar expected yield but greater variability, but they become risk-seeking, that is switch to a preference for the high variance option, when their body weight or blood sugar is so low as to promise overnight starvation and death unless food can be found at a higher than average rate (Caraco, Martindale & Whittam 1980).

In the EEA, it's unlikely that singular, nonflexible solutions to particular adaptive problems would have been successful for all individuals. For example, suppose we hypothesise that, in the EEA, it would have been reproductively advantageous for males to have a preference for casual sexual encounters. It seems relatively clear that various factors could moderate this strategy's success. For example, unattractive males might have experienced significantly less success in securing short-term mates than did attractive males. Thus, in the EEA, an unattractive male might have been reproductively better off pursuing long-term mates, and attractive males reproductively better off pursuing short-term mates. In this case, therefore, selection might have designed mating adaptations to feature both strategies, to be contingently triggered.

So instead of being blind, inflexible instincts, psychological adaptations can be thought of as, and are likely to be, conditional strategies. They can be conceptualised as being facultative responses, as generating a variety of behaviours, depending on what environmental cues are being received. Diverse inputs, diverse outputs. So, for example, it's not the case that 'we're risk-averse'—which, after all, strictly speaking false, as attitudes to risk vary greatly between individuals and within an individual's lifetime. Rather, it's under conditions *X1... Xn,* risk averse behaviour will be generated; under alternative conditions, risk friendly behaviour will more likely obtain.

Buss and Greiling (1999) ask us to consider a man who is married to a woman whose value in the dating market is higher than his own. Would the misalignment in mating value influence the calibration of the man's jealous sensitivities? Adaptationist considerations suggest it would: 'a mechanism for adjusting one's threshold for jealousy would have resulted from thousands of selective events in the evolutionary past in which a mate value discrepancy, on average, was associated with a greater likelihood of a partner's infidelity or defection.' (Buss and Greiling, 1999: 220) Hence, his relationship with his wife might lower his threshold for jealousy than it would otherwise have been if his perceived market value were equal to, or higher than, hers.

So although evolutionary psychologists postulate that psychological adaptations are species-typical ('universal') at the level of design, these species-typical psychological adaptations can generate a rich diversity of behaviours within populations and between populations—the kind of behavioural diversity we observe around us. 'For this reason, much of the study of behavioral variation can be recast as the study of the underlying (and usually) universal psychological adaptations that generate variation in response to circumstantial input' (Cosmides and Tooby, 1995: 17). Indeed, Tooby and Cosmides (1992) coined the term 'evoked culture' to refer to the fact that diverse inputs evoke different behavioural repertoires, thereby forging different elements of culture.

Note that we don't need to attempt to account for all behavioural diversity in terms of different cues triggering different conditional outputs. For example, genetic variation within and between populations could lead to biases in the calibration, and therefore the behavioural output, of our species-typical psychological adaptations. As we shall especially see in the next chapter, because evolutionary psychology research focuses on the design of species-typical psychological mechanisms, and not genetic variation within and between populations, its hypotheses (unsurprisingly) tend not to include such considerations. But, as shall become clear in the next and subsequent chapters, we need not see evolutionary psychology hypotheses as one-shot events and hypotheses can be progressively modified over time to incorporate findings from cognate and adjacent fields such as behavioural genetics.

2.4 Developmental Complexity and Robustness

Adaptations are heritable and reliably developing traits. Developmental outcomes are the result of processes in which there is an ongoing linkage between genes and influences from outside the genome and organism.

Developmental systems theory, pioneered by Susan Oyama, Paul Griffiths and Russell Gray, is a conceptualisation of development that seeks to capture and do justice to the sophisticated and complex interrelations between resources that give rise to traits. Developmental systems theory 'is not a theory in the sense of a specific model that produces predictions to be tested against rival models. Instead, it is a general theoretical perspective on development, heredity and evolution' (Oyama *et al.*, 2001: 1–2).

Sterelny states that most developmental system theorists, who are a 'loose and evolving alliance', would subscribe to the following claims: '(1) genes are just one element of inherited developmental resources; (2) they are but one critical element in the developmental matrix responsible for ontogeny; and (3) organisms in part construct their world, as well as adapt to it' (2007: 180).

The first claim relates to dethroning genes in inheritance by adopting an extended view of inheritance. Intergenerational similarity is sustained by a 'matrix of interacting resources', including not just genes but also things like symbiotic microorganisms and culturally transmitted information (*ibid.*: 178). The second claim is concerned with dethroning genes in development. More precisely, in developmental systems theory genes are not accorded a privileged causal role in the explanation of the development of an organism's phenotype. Instead, a 'causal democracy' is recognised: 'Not only is most standard interactionism shot through with asymmetries, but the notions of causal *symmetry*, or *parity*, which do have a democratic ring, inform the very concept of a developmental system' (Oyama, 2001: 183; original emphasis) As Sterelny (2007) notes, while every developmental perspective is 'moderately interactionist' in that both genetic and environmental contributions to an organism's phenotype are stressed, developmental systems theory is 'radically interactionist' in that it challenges dichotomous accounts of development that partition causal factors into opposing sides of a boxing ring, genes in the red corner, environment in the blue corner. The third claim is known as 'niche construction', the process in which organisms alter their environments. Again, the idea here is to emphasise the active role of something that is usually thought of as being passive.

Evolutionary psychologists can agree with much of the basic picture, the basic holism, underlying developmental systems theory—and often they do. When discussing development from an adaptationist perspective, Tooby and Cosmides (1992: 83–84) write,

> By changing either the genes or the environment any outcome can be changed, so the interaction of the two is always part of every complete explanation of any human phenomenon. As with all interactions, the product simply cannot be sensibly analyzed into genetically determined and environmentally determined components or degrees of influence... 'Biology' cannot be segregated off into some traits and not others.

Indeed, Tooby and Cosmides (1992: 84) even speak of 'developmentally relevant environments', which sound strikingly like 'developmental systems': 'It is this developmentally relevant environment—the environment as interacted with by the organism—that, in a meaningful sense, can be said to be the product of evolution, evolving in tandem with the organism's organized response to it'. Hence, 'The critical question is not... whether every human male in every culture engages in jealous behaviors ... instead, the most illuminating question is whether every human male comes endowed with developmental programs that are designed to assemble (either conditionally or regardless of normal environmental variation) evolutionarily designed sexual jealousy mechanisms' (*ibid.*: 45).

This honouring of causal holism is not restricted to Tooby and Cosmides. Buss (1995: 5) writes,

> The key issues of this debate have been obscured by false dichotomies that must be jettisoned before we can think clearly about the issues—false dichotomies such as 'nature versus nurture,' 'genetic versus environmental,' 'cultural versus biological,' and 'innate versus learned'. These dichotomies imply the existence of two separate classes of causes, the relative importance of which can be evaluated quantitatively. Evolutionary psychology rejects these false dichotomies.

Symons (1992: 140) concurs: 'every part of every organism emerges only via interactions among genes, gene products, and myriad environmental phenomena'. And Bjorklund and Bering (2002: 6) stress that 'individual differences in educability should not be viewed as "genetic" or "environmental" in nature, but rather as the result of the transaction between multiple levels of organization'.

So evolutionary psychology is an arena where the old antagonisms no longer pertain. If we claim that *X* is an adaptation for *Y*, we should be as committed to saying it's socially constructed, a product of post-industrial society, as we would be to saying it has a genetic basis.

Yet there is a common belief that developmental considerations discredit evolutionary psychology. One source of this belief is the misconception of evolutionary psychology focusing on genes. The following paragraph could appear without much challenge in almost any sceptical paper and book:

> it's increasingly apparent that development plays a crucial role in explaining human nature. The old evolutionary psychology picture was that a small set of genes—a 'module'—was directly responsible for some particular pattern of adult behavior. But there's more and more evidence that genes are just the first step in complex developmental sequences, cascades of interactions between organism and environment, and that those developmental processes shape the adult brain. Even small changes in developmental timing can lead to big changes in who we become (Gopnik, 2013: 323).

I'm not convinced that any leading evolutionary psychologist ever endorsed the 'old picture' or rejects the (presumably) 'new' one. Note the invocation of the word 'old', the implication that evolutionary psychology is outdated. We'll return to this theme in the final chapter.

Other sceptics voice a similar thought: Lickliter and Honeycutt (2003: 820) claim that 'the preconceptions of evolutionary psychology ... center around the assumption that basic aspects of an organism, including its morphology, physiology, and psychology, are best understood as the products of its genes'. Nonsense: we've already seen evolutionary psychology is committed to environments playing a complex and indispensible role in every step of the causal chain (see also Dickins and Dickins 2008). Lickliter and Honeycutt (2003: 821) further claim that evolutionary psychology views the environment as 'secondary to the role of genetic factors'. Again, this is not the case: 'evolutionary psychologists do not partition genes and environment into primary and secondary roles' (Buss and Reeve, 2003: 851).

When speaking of evolutionary psychologists postulating specialised psychological adaptations or modules, Heyes (2012: 2094) claims that 'Experience was assumed to play a limited role in the development of these modules'. Shea (2012: 2234) 'rejects an Evolutionary Psychology that is committed to innate domain-specific psychological mechanisms: gene-based adaptations that are unlearnt...'. But again, this simply cannot be the case. Experience is essential during critical developmental stages, pre-natal and post-natal. For example, there appears to be a critical period for language development: successful language acquisition must occur during the period from early childhood to puberty; language learning after this critical period will be increasingly difficult and less successful (Lenneberg, 1967). Pinker (1994) illustrates this critical period with the example of 'Chelsea'. Chelsea was born deaf, but was misdiagnosed as retarded. In her early thirties she was correctly diagnosed as deaf and she received hearing aids and intensive language instruction. However, despite now hearing, her subsequent language acquisition was severely compromised: she produces strings of words, not proper sentences.

Another misconception is to hold that the complex interactions between genes and environment throughout development problematises the notion of developmentally robust adaptations—that if we take seriously the multiple causal factors in the development of an organism, then it's unlikely developmentally robust psychological adaptations would arise. But this leads to an absurdity: if sophisticated interactionism destabilises robust adaptations, we wouldn't have physiological adaptations!

Physiological and psychological adaptations are reliably developmental outcomes because of, *inter alia*, reliably recurring environmental features. Carruthers stresses that 'natural selection can rely on the presence of reliably recurring features of the environment (such as the presence of face-shaped stimuli) when selecting for particular developmental programs' (2006: 162-3). Barrett (2007a: 188) articulates the point perfectly:

> Reliably developing aspects of the phenotype often have the appearance of 'innateness' in some respects. They are produced whenever the developing individual's environment sufficiently matches the ancestral one along relevant dimensions. For example, developmental schedules of some skills, such as the ability to distinguish between animates and inanimates, are relatively invariant across

> very different cultures and environments ... However, this does not tell us what factors in the environment might contribute causally to development of the competence. Innateness in the folk sense of lack of environmental input is not mandated.

This goes back to a point I made earlier—that the EEA of adaptations can often match today's environment. Despite the diversity found in Japanese, Brazilian, Nigerian, German societies and so on, they still share strikingly similar ecological, developmental and societal features. These invariant features enable a person to usually develop ten fingers, a pair of eyes, a pair of legs and so on, regardless of where he or she is born. Likewise for psychological adaptations: a woman born in Nigeria will tend to develop the same set of psychological adaptations as a woman in born in Brazil. Whether she is born in Nigeria or Brazil, she will still receive parental investment; she will be raised in a language community; she will encounter animate and inanimate objects; she will learn social norms; she will need to navigate a complexity of social challenges; she will still read human faces; she will engage in courtship and mating; and so on. Many of the reliably developing psychological adaptations that arise from the complex interactions between genes and environment will seem 'innate' in a folk biological sense, in the sense of not requiring environmental inputs, but this is only an appearance, one resonating with our folk intuitions, and shouldn't be construed as a scientific claim.

So it's a misconception to think that the complexity of the interactions between genes and environments discredits, or is in tension with, the notion of developmentally robust psychological adaptations. Indeed, Tooby and Cosmides' repeated use of the term 'developmentally relevant environment' should unnerve those who insist on problematising psychological adaptations with developmental complexity. As Dennett (2011: 483) nicely puts it, 'the oft-implied claim is that one should be an evo-devo theorist *instead of* an adaptationist/selectionist/agentialist. Why not be both?' (original emphasis).

2.5 Psychological Adaptations as Modules

There is some confusion about the concept of modularity—especially among sceptics. This confusion arises from the fact that the term 'module' has developed a range of meanings and is often used without making clear which meaning is intended. Like many concepts, the concept of modularity means different things to different people or, as Ermer *et al.* put it, 'different research communities' (2007: 153).

Fodor (1983) characterises modularity in terms of nine properties. In order of Fodor's original presentation, modules are information processing systems that are (1) domain-specific (they process a restricted range of inputs); (2) mandatory (they operate in an automatic way once the domain-specific stimuli is present); (3) inaccessible (higher levels of cognitive processing have no access to the internal processes of a module); (4) fast (they generate outputs quickly); (5) informationally encapsulated (modules can't draw on information held outside of it); (6) shallow (they have relatively simple outputs); (7) localised (they are realised in a fixed, dedicated neural architecture); (8) subject to patterned breakdowns; and (9) their ontogeny have a characteristic pace and sequence.

For Fodor, the key feature of a module is the fifth property, namely informational encapsulation. Fodor's definitional emphasis on informational encapsulation arises from his emphasis on a particular kind of brain system: perceptual systems. Perceptual systems are tasked with a particular function: to interpret stimuli rapidly and correctly. Fodor reasoned that in order to achieve this function, the input systems need to operate automatically and not be slowed down by the operation of other systems. This kind of system is essentially 'hardwired pipelines bringing information to central systems' (Barrett, 2007b: 162). If one is operating under such a narrow definition of modularity, then it 'becomes true virtually by definition' that most of the mind is not modular (*ibid.*: 163).

Crucially, however, few brain systems are like this. For one, brain systems are thickly interconnected (*ibid.*: 163). Indeed, Fodor recognised this, arguing that there are only a handful of modules. Carruthers notes that 'some of the items in Fodor's list will need to get struck out as soon as we move to endorse any sort of central-systems modularity, let alone entertain the idea of *massive* modularity' (2006: 5, original emphasis). In particular, in order to have modularity in central process cognition, informational encapsulation, the most important feature in Fodor's understanding of modularity, needs to be jettisoned.

But if informational encapsulation is not the key to understanding modularity, what is? For evolutionary psychologists, 'functional specificity' is the fundamental feature of modularity (Barrett, 2007b). Crucially, they argue that brain systems can be functionally specialised without being informationally isolated and encapsulated (*ibid.*: 163).

Barrett emphasises that 'the core of specialization in biological systems is the fit between *form* and *function*' (*ibid.*: 164, original emphasis). Different functions demand different design criteria. Whereas perceptual systems need to exclude information in order to execute their function, systems responsible for mate choice need to integrate information from a variety of sources to achieve their function (*ibid.*). Both systems are modular in the sense of being functionally specialised but they radically differ in their design criteria. It can therefore be a 'major mistake' to apply 'design criteria from one kind of system to another' (*ibid.*: 167). Accordingly, as Fodor's concept of a module relates to the design criteria of an uncommon kind of functionally specialised processing structure, it 'is neither useful nor important for evolutionary psychologists' (Ermer *et al.*, 2007: 153).

Despite these statements, some opponents of evolutionary psychology claim that one of the main characteristics of a module as understood by evolutionary psychologists is informational encapsulation (see, e.g., Buller, 2005: 128). They then proceed to point out how this property is implausible for many proposed psychological adaptations. The dialectic usually runs along the lines of 'the brain is plastic, but modules are rigid'. As Buller and Hardcastle (2000: 311) put it:

> Brain plasticity belies the idea of encapsulated modularity, for our information processing streams are not really separate streams at all. There is much informational overlap between what are normally thought of as distinct processing areas. In other words, whatever modules one might want to identify in the brain are not as distinct, or informationally encapsulated, as evolutionary psychologists typically imply.

But, as we have seen, evolutionary psychologists don't adopt this conception of modularity, hence the argument is irrelevant. Barrett (2006: 201) observes that:

> evolutionary psychologists have stressed repeatedly that the core of their notion of modularity is functional specialization (Barrett, 2005b; Pinker, 1997; Tooby & Cosmides, 1992). However, others have read modularity claims as implying much more: for example, automaticity (DeSteno et al., 2002), encapsulation (Fodor, 2000), localization (Uttal, 2001), lack of plasticity (Buller & Hardcastle, 2000; Elman et al., 1996; Karmiloff-Smith, 1992), and innateness verging on preformationism.

So, some of the widespread controversy surrounding the concept of modularity in evolutionary psychology is rooted in semantic confusion.

It might be fruitful, therefore, for evolutionary psychologists to simply drop the label 'module' while retaining the concept of specialised information processing. Indeed, one might be inclined to think that Cosmides and Tooby have caused the semantic confusion by failing to be more explicit with how their understanding of modularity differed from Fodor's.

However, Cosmides and Tooby blame Fodor for the semantic confusion: 'Fodor uses some terms differently from the way we do, leading to some considerable confusion in the literature' (Tooby *et al.*, 2005: 309, n. 4). They claim that their understanding of modularity has a longer pedigree than Fodor's. They point out that the concept of module first arose in artificial intelligence to simply refer to the concept of 'a mechanism or program that is organized to perform a particular function' and charge Fodor with clouding this simple meaning 'in favor of an eccentric set of criteria' and thereby sowing 'a great deal of confusion' over the term. 'For evolutionary psychologists, the original sense of module—a program organized to perform a particular function—is the correct one' (Ermer *et al.*, 2007: 153).

Carruthers (2008) adds that most philosophers suffer from what he calls 'Fodor-Fixation', labouring under a fable about modularity that runs as follows: Fodor engineered a notion of modularity fit for service in cognitive science, with the implication that only peripheral systems are modular; accordingly, theorists who wish to advance the thesis of massive modularity must weaken the concept of modularity in a way that still preserves the central idea that modules are encapsulated processing systems. This fable, claims Carruthers, obscures philosophers' understanding of the conception of modularity operational in evolutionary psychology.

However, other misunderstandings over modularity in evolutionary psychology might also be due to what Carruthers calls the 'highly misleading' language evolutionary psychologists themselves are sometimes prone to use (2006: 21). Carruthers stresses the importance of not contemplating modules as streamlined pipes. Using the figurative language of 'Swiss army knives' and 'adaptive toolboxes', evolutionary psychologists perhaps unwittingly fuel this misapprehension.

Modules are 'distinct functionally specialized cognitive systems, that is all' (Carruthers, 2006: 21). Each module will have a 'distinct neural realization' but the neural realisation might 'be dispersed across a number of different brain regions' (*ibid.*: 62). One of the virtues of a module is 'the flexibility of its demand for real estate' (Pinker, 1997b: 30). Modules 'are biological systems, and like most such systems they are likely to be built by co-opting and connecting in novel ways resources that were antecedently available in the service of other functions' (Carruthers, 2006: 21) This 'recruiting and cobbling together in quite inelegant ways resources that existed antecedently' leads to '"kludgy" architectures that look decidedly awkward in design terms' (*ibid.*: 62). They look 'more like roadkill, sprawling messily over the bulges and crevasses of the brain' (Pinker, 1997b: 30). Hence, modules are not 'elegantly engineered atomic entities with simple and streamlined internal structures' (Carruthers, 2006: 62).

Carruthers (2008: 294) provides a succinct summary of the concept of modularity in evolutionary psychology:

> What emerges from these arguments is a notion of 'modularity' according to which modules are function-specific processing systems which exist and operate independently of most others, and which have complex, but limited, input and output connections with others. Each of these systems will have a distinct neural realization, and will be frugal in its use of information, while having internal operations that are inaccessible to others.

All modules will have the properties just cited. Furthermore, 'almost all' will be domain specific and 'most' will be 'innate'. 'Some' modules will even be encapsulated (*ibid.*: 295).

2.6 The Failure to Infer Massive Modularity from Empirical Adaptationism

Some evolutionary psychologists have attempted to infer from general empirical adaptationist considerations that the mind is probably massively modular (whether in the moderate or strong sense). There has been a tremendous volume of discussion on this and a number of arguments have been exchanged in the literature. The consensus, however, appears to be that there is no secure or good inference from empirical adaptationism to the likelihood of massive modularity: domain-specific processes are no more likely than domain-general processes.

A brief look at one kind of argument will suffice to show the indeterminacy of such inferences—which is a theme we'll explore in more detail in a moment. The argument I have in mind is evolutionary psychology's version of the poverty of the stimulus argument. This argument is intended to establish a remarkably weighty conclusion, namely that 'it is in principle impossible for a human psychology that contained nothing but domain-general mechanisms to have evolved' (Cosmides and Tooby, 1994: 90). The argument roughly runs as follows. Suppose humans only had domain-general mechanisms. Accordingly, the information available to them will be dramatically limited to only what can be gained from perception, which is impoverished, too impoverished to enable humans to survive and reproduce. Why is that? Well simply because there would have been insufficient time or available information for any human to solve successfully from scratch the multitude of adaptive problems they would face. So domain-specific processes would more likely have been selected for and as our ancestors faced a large range of problems we have a large range of domain-specific processes.

Leaving aside problems with the poverty of the stimulus arguments in general, Samuels (1998), Fodor (2000) and Buller (2005) have responded to this argument by highlighting that evolutionary psychologists have conflated two distinct concepts: the concept of modularity with the concept of innateness. As we saw in the previous chapter, all that is required for a mechanism to be domain-general is that it is not domain-specific. In other words, that is not dedicated to solving problems in a specific cognitive domain. Therefore, in order to even put forward this argument, one must assume that a mind that consists of only a small number of domain-general mechanisms is a mind that must contain no innate knowledge. But why is that? It's not a conceptual truth: it is entirely conceivable for a domain-general mechanism to possess innate knowledge. Indeed, Richard Samuels (1998) postulates what he calls the 'library model of cognition' in which there is domain-general processing with domain-specific knowledge. Fodor (2000) points out that the concepts of innateness and modularity are actually 'independent in both directions':

> you can thus have perfectly general learning mechanisms that are born knowing a lot, and you can have fully encapsulated mechanisms (e.g. reflexes) that are literally present at birth, but that don't know about anything except what proximal stimulus to respond to and what proximal response to make to it (*ibid.*: 69).

Hence, even if we can infer from general empirical adaptationist considerations that it's likely our mind will contain innate knowledge, it is equally as likely that this would result in the evolution of cognitive architecture consisting of only a few domain-general mechanisms with the required innate knowledge.

So even if we assume empirical adaptationism, the trouble with running *a priori* arguments for massive modularity is that there is no single privileged evolutionary prediction as to what kind of cognitive architecture we have. It seems clear that a cognitive architecture constructed with many specialised adaptations is just as plausible *a priori* as one constructed with a few general adaptations. As Lewens puts it, 'perhaps we responded to those problems not by acquiring a set of rigid adaptations, but by acquiring abilities to respond in ad hoc ways to new problems as they arose' (2007: 153). Although I disagree with the characterisation of specialised psychological adaptations as being 'rigid'—they are anything but, as we have seen—the general point is taken.

2.7 Methodological Objections

Even if considerations of selection do not guarantee massive modularity in any shape, the possibility remains that there might nevertheless be psychological adaptations and that, accordingly, it might be useful to hypothesise between adaptive problems and possible adaptive solutions. But even this has been challenged. The following objections are activated at different points during the methodological process of hypothesising between adaptive problem and adaptive solutions. And, as we will see, they are legitimate and pressing questions—questions that many have taken to have sufficiently destabilised methodological adaptationism in psychology.

2.7.1 No Stable Problems Objection

Were adaptive problems sufficiently stable to allow selection to engineer stable solutions? What if the relationship between environments, especially social environments that involve interdependent decision problems, and evolving populations is far more dynamic than evolutionary psychologists realise?

Sterelny (1995: 372) claims that there are, in fact, 'no stable problems to which natural selection can grind out a solution'. This argument also appears in Sterelny and Griffiths (1999). As you can see, this kind of argument, if successful, would pull the entire rug from underneath evolutionary psychology. The thought is that evolutionary arm races destabilise adaptive problems. And as evolutionary psychology trades on adaptive problems, evolutionary arm races also destabilises evolutionary psychology.

An arms race involves feedback: a problem arises, a solution emerges, but the emergence of the solution creates pressure on the original problem, changes it, transforms it, thereby requiring the solution to change, and so on. Sterelny observes that, 'As men evolved to detect ovulation, women evolve to conceal it. As we evolve to detect cheaters and others of uncooperative dispositions, emotion-mimics evolve better and better fakes of a trustworthy and honest face' (1995: 372). So evolutionary psychology methodology seems to trade on an overly simplistic characterisation of evolution, on stable adaptive problems and adaptive problems. How then can that methodology possibly capture this dynamic, this interactive character of evolution?

Do evolutionary arms races destabilise adaptive problems? I don't believe they do. But I shall leave that issue for the next chapter because it's important to continue building the strongest possible case against methodological adaptationism in psychology.

2.7.2 The Fine Grain Problem

So let's assume there are adaptive problems. How coarse are those problems? How do we individuate adaptive problems? How do we characterise them? How do we individuate them?

The problem of how to individuate correctly domains that specialised adaptations are purported to operate on is what Sterelny and Griffiths (1999) call the 'grain problem'. How coarse or fine is the grain of a domain? How specific is the adaptive problem? Sterelny and Griffiths (1999: 328) ask us to consider the domain of 'mate selection':

> Is the problem of mate choice a single problem or a mosaic of many distinct problems? These problems might include: When should I be unfaithful to my usual partner? When should I desert my old partner? When should I help my sibs find a partner? When and how should I punish infidelity?

The correct identification and characterisation of adaptive problems is not obvious. Is 'mate selection' one adaptive problem, with several sub-problems? Or is 'mate selection' a reference to many distinctive adaptive problems? There seems to be no principled way of answering this. It seems arbitrary.

2.7.3 No Constraints Objection

But suppose we have fixed the grain of the problem according to our own satisfaction. We can now reason from the adaptive problem to the adaptive solution—or vice versa. But how constrained will our reasoning be? How elastic? In other words, from an adaptive problem can we straightforwardly read off an adaptive solution (or vice versa)? Or will there be many possible reasoning trajectories—perhaps too many?

A long standing, deeply entrenched worry in the literature is that the reasoning between adaptive problem and adaptive solution is too elastic. Fitness considerations can be so flexible that it seems possible to reason between adaptive problem *A1* and a spectrum of possible adaptive solutions *S1, S2, ... Sn*. And so too the other way: with sufficient ingenuity, fitness benefits can be imagined for almost any observed behaviour.

This concern is frequently cashed out into a very strong position: that evolutionary psychology hypotheses are unconstrained, that there is a free for all, a Darwinian Wild West of hypotheses. For example, Richardson declares that 'Just about anything is consistent with some evolutionary model or other' (2007: 65). Gray *et al.* (2003) approvingly quote Rosen (1982), who quips that there are only two limiting factors that constrain adaptation hypotheses: the imagination of the theorist and the gullibility of the audience.

So there is indeterminacy with respect to both the grain of the adaptive problem and, more seriously, in the (potentially runaway) multiplicity of possible hypotheses. But let us suppose that we have settled upon a hypothesis. Suppose, despite the indeterminacy, the elasticity, we have arrived at a hypothesis: that adaptive problem *A1* is solved by psychological adaptation *S1*. How good would that be as an explanation?

2.7.4 The Thinness of Evolutionary Psychology Explanation

Richardson (2007) claims that evolutionary psychology explanations are actually nothing more than 'crude speculations' (*ibid.*: 96), 'unconstrained speculation' (*ibid.*: 171), 'empty generalisations' (*ibid.*: 171), and that we should 'reject' them as such (*ibid.*: 38). Richardson does not deny that our psychological systems are the product of a long history of evolution, and that some of these might well be adaptations. Furthermore, Richardson is not claiming that evolutionary psychology hypotheses are false, nor does he offer alternative evolutionary scenarios for their hypotheses (*ibid.*: 139). His primary contention, the gravity of his complaint, is that as explanations they're evidentially unsupported—and unlikely to ever be supported.

Richardson's strategy is straightforward. First, establish what a complete adaptationist explanation looks like. Second, argue that evolutionary psychology hypotheses fall far short of this standard. Third, suggest that evolutionary psychology will never be able to produce the evidence required to establish complete explanations. From this, Richardson believes, evolutionary psychology hypotheses can be declared as 'little more than storytelling' (*ibid.*: 147), 'idle Darwinizing' (*ibid.*: 183), valuable only in discrediting the method that produced them.

Richardson points out, quite rightly, that adaptation claims are historical claims: to claim that *X* is an adaptation is to make an historical claim. 'Whether a trait is an adaptation thus depends on its evolutionary history; and explaining some trait as an adaptation depends on knowing the evolutionary history that produced it' (2007: 98). A complete adaptationist explanation cannot simply indicate a scenario of how selection might have favoured the evolution of some given trait; it must also specify the history of the trait's selection, reliable historical information concerning under what conditions selection actually occurred. To vindicate such historical claims, we need evidence concerning variation in ancestral populations, evidence concerning their heritability and the like. Richardson draws on Brandon (1990), who lists five categories of evidence complete adaptation explanations should provide: (1) selection, (2) ecological factors, (3) heritability, (4) population structure and (5) trait polarity.

First, there must be evidence that selection has, in fact, occurred; without such evidence alternative evolutionary explanations are viable. As mentioned in Chapter 1, without variation there can be no selection, so we need information about trait variation in ancestral populations, as well as the relative rates of survivorship and reproduction of organisms with varying traits—that is, information on differential fitness. Second, the ecological factors and pressures, whether abiotic or biotic, determining relative survivorship and reproduction should be specified. For example, if a specific predation factor affected relative survivorship and reproduction, this should be specified. Third, there must be evidence that the trait in question is heritable. Without heritability, selection is powerless. Fourth, as the strength and direction of selection crucially depends on population size, gene flow, interbreeding and mutation rates, there must be reliable information regarding population structure. Different population structures will be affected by selection in different ways. Richardson points out that gene flow can reduce the impact of selection in a local population, and that small populations are more sensitive than larger populations to chance. Fifth, phylogenetic information concerning primitive and derived traits is required. Is the alleged adaptation a primitive trait or a derived trait? Is the trait ancestral within a clade or is its presence the result of convergence among lineages? Answering this requires an independently established phylogeny.

Richardson applies Brandon's criteria to Buss's research on the evolution of sex differences in jealousy (2007: 60–4), Pinker's research on the evolution of language (*ibid.*: 124–9), and Cosmides and Tooby's research on their proposed cheater detection mechanism (*ibid.*: 132–6). In his judgement, evolutionary psychologists frequently and dismally fail to provide the evidential particulars, the actual selectionist history, required of adaptation explanations. For example, 'In Buss's exposition of the case for jealousy, there is no significant appeal to the historical conditions of human evolution, aside from general appears to the conditions of the Pleistocene' (*ibid.*:88). Little detail is cited about the structures of ancestral populations, about immigration and gene flow. Few details are provided as to the variation of traits between organisms, the degree of selective advantage, the ecological mechanisms producing selection, or the trait's degree of heritability. And comparative studies and phylogenetic analyses are seldom, if ever, cited or performed.

No matter how plausible evolutionary psychology explanations might sound, they lack the crucial kinds of information concerning the selective pressures and historical contingencies shaping our psychology.

> It is, of course, plausible that our ancestors engaged in sharing of food and other resources—so, for that matter, do chimpanzees—and it may be true that there was considerable variance in food availability. But even if true, these general facts and constraints are inadequate to capture the kind of historical information we would need to construct an evolutionary explanation, including factors such as the kind and extent of variation present in our ancestors, the actual environmental features that affect survival and reproduction, and demographic factors (*ibid.*: 138).

Although Richardson only considers a few examples, we can grant, I believe without controversy, his claim that evolutionary psychology explanations frequently don't provide the full range of evidential particulars expected of complete explanations. When evolutionary psychology research for some given phenomenon is presented as an explanation for that phenomenon, it's a safe bet we can identify missing details. Indeed, highlighting the incompleteness of evolutionary psychology explanations is a standard manoeuvre in the sceptical literature. For example, Profet (1992) hypothesised that pregnancy sickness is an adaptation to avoid toxins harmful for the fetus. Sterelny (1995) pointed out that the pregnancy sickness might well have benefits but that Profet didn't balance the purported benefits against possible costs, to perform a proper cost–benefit analysis. Evolutionary psychologists are also frequently berated for not adopting comparative tests for traits across a range of species closely related to us. 'Evolutionary Psychology sometimes gives the impression that new cognitive processes appeared suddenly and fully-formed as a result of lucky genetic mutations and fierce, unimodal selection pressures' (Heyes, 2012: 2093).

Furthermore, Richardson claims are not only evolutionary psychology explanations unsupported as they currently stand, but they're also unlikely to be supported in the final verdict. Unfortunately, Richardson doesn't provide an argument for this, and admits as much: 'I suggest, but do not demonstrate, that in the end we are unlikely ever to have the sort of evidence that would be required to make it reasonable to embrace the hypotheses of evolutionary psychology' (2007: 38). Beyond one or two statements expressing this pessimism, he doesn't push the thought any further.

As evolutionary psychology research stands woefully deficient, in Richardson's judgement this means evolutionary psychology simply consists of 'empty generalisations' and 'unconstrained speculation' (2007: 171)—just so stories of what could have been, speculation dressed up in scientific jargon. Just so stories like Kipling's *Just So Stories for Little Children*, imaginative, fanciful origin stories for the distinctive traits various animals have. How did the camel get its hump? The hump allows the camel to work several days without eating. And as you can imagine perhaps long ago the camel refused to work and so was given the hump by a djinn as punishment. The adaptationist explanations offered by evolutionary psychologists are 'Just So Stories for Adults'—perhaps a little less fanciful, perhaps a little more plausible, but still just so stories. Evolutionary psychologists might be inclined to establish their scientific credentials by off-hand references to evolutionary theories and talking about adaptive problems, but on the whole their hypotheses are evidentially weak when judged against the criteria above. This is scientifically indefensible.

Richardson (2007) is but the latest formulation of the complaint. It might not be the last. This line of thought has a long pedigree, going back to Lewontin and Gould. Lewontin declares 'we know essentially nothing about the evolution of our cognitive capabilities and there is a strong possibility that we will never know much about it' (1990: 229). Unsurprisingly, Gould concurs:

> how can we possibly know in detail what small bands of hunter-gatherers did in Africa two million years ago? ... how can we possibly obtain the key information that would be required to show the validity of adaptive tales about an EEA: relations of kinship, social structures and sizes of groups, different activities of males and females, the roles of religion, symbolizing, storytelling, and a hundred other central aspects of human life that cannot be traced in fossils? We do not even know the original environment of our ancestors—did ancestral humans stay in one region or move about (2000: 120)?

Kitcher and Vickers also focus on evolutionary psychology research lacking evidential particulars, claiming that the historical information they put forward is rudimentary, that the underlying genetic details are absent, and conclude 'we ought to dismiss these suggestions as vague speculation' (2003: 337).

So it appears that this indeterminate or rickety method produces thin 'how possibly' explanations rather than deep 'how actually' explanations, insufficiently determinate and insufficiently detailed to be compelling as complete explanations.

All of this throws the method's explanatory value into serious question. But the situation actually gets worse: underdetermination of theory by evidence means that whatever explanation is put forward to account for some phenomenon, an alternative explanation equally as fitting the phenomenon can also be postulated. Indeed, even though evolutionary psychologists conduct experiments to see whether people have psychological adaptations, these results are still subject to underdetermination. Furthermore, one can always challenge the interpretation of experimental results or even the reliability of the experiment. Buller (2005) has especially leveraged such manoeuvres against evolutionary psychology. However, it's important to realise these considerations are not unique to evolutionary psychology—they are ubiquitous to any scientific enterprise. Any principle governing theory selection in light of underdetermination is a principle that can be applied to this debate. I shall make further remarks on this in Chapter 5.

As you can expect, these kinds of considerations have led to a near consensus in philosophy that the adaptationist method in psychology is explanatorily weak. But what about its purported heuristic value?

2.7.5 The Mystery of Discovery

Many philosophers have entirely or almost entirely ignored evolutionary psychology's purported heuristic dimension. And it's not difficult to see why. We have seen that fitness thinking represents insufficient constraint on adaptationist methodology. How can evolutionary psychology possibly be heuristic if fitness considerations between adaptive problem and adaptive solution can yield multiple trajectories? If I consider adaptive problem *P1* and fitness thinking generates a long list of possible adaptive solutions *S1, S2, ... Sn*, how can adaptationist methodology be of predictive value? How can adaptationist methodology discover anything new?

Indeed, while it seems clear what an adaptationist explanation looks like and precisely why reasoning between adaptive problem and adaptive solution can't possibly produce the assortment of evidence needed for a complete adaptationist explanation, it's somewhat unclear in the literature in virtue of what could a methodology be considered heuristic and whether evolutionary psychology methodology can be considered as such. This way of seeing evolutionary psychology is undeveloped in the philosophy literature. But we will develop it further in the next chapter.

For now it is sufficient to note that, as we saw in the previous chapter, evolutionary psychologists claim to have made important discoveries in psychology—the key example being the cheater detection mechanism. What are we to make of such claims?

Unlike many others, Schulz (2011) has taken a closer look at some of the purported trailblazing examples of evolutionary psychology as a heuristic. And upon closer examination, he has found something apparently destabilising to evolutionary psychology's heuristic credentials: these trailblazing exams are actually explanatory, not heuristic. If this is indeed the case, and if indeed this is representative, this further illustrates the method's limited usefulness, perhaps even its bankruptcy. Note, unlike Buller (2005), Schulz (2011) is not concerned with the finer details of confirmation, whether such and such experiment actually supports a given hypothesis—just whether the hypothesis is genuinely novel in the first place or whether it is instead actually (attempting) explanation or accommodation of what was already known.

Schulz examined the cheater detection hypothesis. Recall from the previous chapter that the cheater detection hypothesis is often submitted as being the leading example of a successful hypothesis generated by reasoning from the past to the present. Cosmides and Tooby (1992) hypothesised that selection would favour the evolution of a mechanism dedicated to detecting, and avoiding future cooperative effort with, cheaters—a cheater detection mechanism, a psychological adaptation for representing and computing social exchanges. They found evidence for this hypothesis in the Wason selection task.

Sounds impressive. Here we have a case of forward engineering: beginning with a selectionist scenario, postulating a hitherto unknown psychological mechanism, and then canvassing evidence to vouch for it. Or do we? Although this is how the hypothesis is typically described as being generated, the description is problematic. Wason selection task anomalies were discovered and known before the cheater detection mechanism was hypothesised: the original Wason selection task was published in 1966 and the Griggs and Cox version was published in 1982. Of course, the fact that Wason selection task anomalies were known before the cheater detection hypothesis was formulated doesn't in itself prevent the Wason selection task anomalies being scored as a novel prediction of the cheater detection hypothesis—a known fact can be considered a novel prediction of a hypothesis as long as it wasn't used in the construction of the hypothesis. But there lies the problem: it seems that the cheater detection hypothesis was constructed specifically to account for Watson selection task anomalies. To see this, merely observe that Cosmides's 1985 PhD thesis, where the cheater detection hypothesis was first proposed, is entitled 'Deduction or Darwinian Algorithms? An *Explanation* of the "Elusive" Content Effect on the Wason Selection Task' (my emphasis).

Schulz is surely right when he says that this is a clear case of evolutionary theory being applied in an explanatory way and not in a heuristic way.

> To see this, note that the key social psychological effect difference to be accounted for had *already been known* when Cosmides & Tooby put their evolutionary hypotheses forward: the difference in the success rates in evaluating the two kinds of conditionals was the *starting point* of their evolutionary investigation—and not an end state (2011: 222; original emphasis).

The obvious response here is to point out how much research has been sparked by the efforts of Cosmides and Tooby. Before Cosmides and Tooby, the Wason selection task was an obscure anomaly, unanchored, being claimed by no theory. Enter Cosmides and Tooby: suddenly researchers take notice of inferential reasoning over conditional rules, both proponents and opponents of evolutionary psychology, and an avalanche of experiments testing competing explanations arises. The Wason selection task was back on the map.

Schulz recognises this response. Nevertheless, he stresses that even though Cosmides and Tooby gave an explanation where none before existed, and even if it led to further tests using the Wason selection task, and even further experiments to rule out competing explanations, this alone doesn't elevate the accommodation into a novel prediction. 'Because of this,' claims Schulz, 'it seems clear that this case does not support a heuristic interpretation of evolutionary psychology—it quite simply does not exemplify any heuristic application of evolutionary theory at all' (*ibid.*: 224).

Having argued that Cosmides and Tooby's cheater detection hypothesis is a product of accommodation, Schulz briefly looks at another well-known example of evolutionary psychology work: Buss's work on 'Sexual Strategies Theory'. However, Schulz judges that this, too, must also be seen as trying to accommodate existing knowledge of the differences in the way men and women choose mates. 'This comes out clearly from the fact that Buss *begins* his research by empirically substantiating the widespread supposition that males tend to want different things from the things that females want (at least in some cases), and then uses Trivers's theory of minimal parental investment to *account* for these differences' (*ibid.*: 226; original emphasis).

Who can argue with that? Disputing the claim that sex differences in attitudes to sex was an idea doing the rounds long before evolutionary psychology entered the scene would be a brave move to make. Schulz further claims that similar judgements can be made about much of Symons's, Daly and Wilson's, and Pinker's work and 'that of many other researchers in this area (for more on this work, see e.g. Barkow et al., 1992)' (*ibid.*: 266). Barkow *et al.* (1992), of course, refers to the seminal *The Adapted Mind: Evolutionary Psychology and the Generation of Culture*, the research programme's first handbook.

Indeed, recall Daly and Wilson (1980), who proposed that we have a psychological adaptation for discriminative parental solicitude. Is it really a discovery that parents generally tend to care less for stepchildren than they do for their genetic children? Even if studies verifying this were originally thin on the ground, and Martin Daly and Margo Wilson performed new studies, wasn't the *Cinderella* Effect already common knowledge? Wasn't this a phenomenon already known, even if not fully established by research, rather than discovered?

The implication is clear; and forgoing discussing other examples, Schulz reaches his conclusion—the contention that most evolutionary psychology practice is accommodation: 'Moreover, it is easy to see that this conclusion generalises to many other evolutionary psychological research projects' (*ibid.*: 226). Hence, while '*merely possible, fictional* evolutionary psychology' is heuristic, '*currently practiced, actual* evolutionary psychology' is not (*ibid.*: 218, original emphasis).

2.8 Metatheory or Marketing?

Psychology clearly hasn't been unified by the emergence of the evolutionary psychology paradigm. And there are no signs of evolutionary psychology's promised revolution. Outside of evolutionary psychology, the multitude of psychology programmes hasn't shown even the slightest sign of re-organising along adaptationist lines.

Leading evolutionary psychologists are not blind to this and they acknowledge the existence of significant opposition to evolutionary psychology. However, there seems to be a belief that this resistance is primarily due to misconceptions and that future generations of social scientists will come to see evolutionary psychology as the key metatheory. Goetz *et al.* paint a triumphant vision of a delayed but upcoming Pax Santa Barbara in psychological science:

> As new psychologists are impartially introduced to EP [evolutionary psychology], as 'traditional' (i.e. anti-evolutionary) psychologists retire, as EP's empirical output grows, as findings from genetics corroborate findings from EP (e.g. Cherkas *et al.*, 2004), as the neural substrates underlying hypothesized psychological mechanisms are indentified (e.g. Platek, Keenan, and Mohamed, 2005 and this volume) and as cross-disciplinary frameworks of evidence are utilized (Schmitt and Pilcher, 2004), EP will emerge as *the* metatheory for psychological science (2009: 15, emphasis original).

But why do the behavioural sciences need to unify? Is there really a need to have a seamless theoretical connection between the various branches of the behavioural sciences and with evolutionary theory?

It's not obvious at all that behavioural science models being mutually inconsistent with one another is a bad thing. Even if many behavioural models are mutually incompatible there isn't really a pressing issue to correct this as long as they're useful. False models can be useful models. Furthermore, even if we accept that the models used in the behavioural sciences should be consistent with evolutionary theory, this is a very weak demand: it doesn't take much effort to make a theory consistent with evolutionary theory. As we saw earlier, radically different cognitive architectures are consistent with evolutionary theory and likewise radically different models can be made consistent with evolutionary theory. Maybe unification might foster innovation? But again this fails to translate into a need to unify. Many would like to see more economists become more empirically grounded, and more sociologists to be more mindful of evolutionary theory: Wouldn't that be enough to secure increased innovation? Why isn't that good enough? Why go for unification across the board?

So the disunity of the behavioural sciences is not an obviously bad thing and so there is no clear need, let alone a pressing need, to unify. Given that the purported rationale for the launch of evolutionary psychology decades ago was to address the disunity of the behavioural sciences, it's somewhat odd that leading evolutionary psychologists haven't really demonstrated that the behavioural sciences even need to be unified by a common framework in the first place. Indeed, quite the contrary, there are good reasons to favour disunity, at least at this point in time.

First, disunity fuels innovation. Many of the fundamental conceptual and methodological debates in the behavioural sciences—methodological individualism versus holism, the level of selection debate in biology and so on—are unresolved and will continue to be unresolved for a long time. Dialectical pressures between rival research agendas give rise to new insights and research: competing schools of thought do battle, their practitioners race in competition with one another, charting new conceptual space and, along the way, making empirical discoveries. Unattached, unconnected mini-theories can also highlight phenomena that other mini-theories or higher-level theories are blind to, leading to new empirical finds. And we're all the better for it. Most behavioural scientists seem quite content with the existence of diverse, conflicting models; perhaps this is a significant reason why.

Second, the possibility and dangers of a limiting or defective meta-theory. Suppose we replace the dazzlingly array of incompatible models with a standardised framework. What if, unbeknownst to us, that framework is defective or limited in some aspect? Adopting a common framework will inevitably lead to the withering away of various types of research that rub against the unifying framework and, accordingly, a common framework with limitations and defects will limit and defect output. Believing that one's conceptual framework is correct is not good enough. Given that so much is still up for grabs in the behavioural sciences, it is short sighted to propose any particular theory as a uniform framework for the behavioural sciences at this point in time.

Indeed, Gintis (2007, 2009) also champions the unification of the behavioural sciences but on a very different basis to evolutionary psychology. Gintis identifies several conceptual frameworks, already in use in different quarters of the behavioural sciences, that when collectively acknowledged and embraced could serve as a unifying framework for the behavioural sciences. The frameworks are: (1) gene–culture coevolution;

(2) the sociopsychological theory of norms; (3) game theory; (4) the beliefs, preferences and constraints (BPC) model; and (5) complexity theory. While applauding Gintis's championing of the unification of the behavioural sciences, Tooby and Cosmides (2007) highlight limitations with the key BPC framework and they, accordingly, re-affirm their belief that evolutionary psychology would be the better metatheory for the behavioural sciences. But, as we have seen, similar concerns about limitations and anomalies can be raised about evolutionary psychology.

Third, the case of the natural sciences. For several decades now physics has been in a state of disunity, with relativity and quantum theories still resisting unification. This situation is likely to continue into the future; and yet, while this is obviously a puzzle and attracts a great deal of interest and research, it's not a pressing issue. Furthermore, we don't blame the disunity in the physical sciences on hostile attitudes or institutional failings, as some evolutionary psychologists seem to do with respect to the disunity of the behavioural sciences.

There is little to no prospect that the behavioural sciences will re-organise along adaptationist lines in the foreseeable future: apart from the failure to *a priori* establish a massively modular mind and the difficulties establishing secure adaptationist knowledge, and apart from the lack of a compelling case to unify and the existence of good reasons to favour disunity, it's simply not obvious or clear how dramatically differently approaches and information gained from those approaches can, in practice, be re-organised along adaptationist lines. As we saw in the previous chapter, leading evolutionary psychologists hold that evolutionary psychology provides the main non-arbitrary way of carving up the mind, providing the true psychology taxonomy. But even if we suppose that is true, and we have now seen there is plenty of room to be sceptical of this and to resist this position, it's entirely unclear how, in practice, the multitude of behavioural science programmes can be connected on that basis—and nor any assurance that such a metatheory will not make invisible what unattached mini-theories make visible. And if evolutionary psychology doesn't seem able to unify the other enterprises immediately adjacent to it in the evolutionary behavioural sciences, such as human behavioural ecology, what hope does it have of unifying the dramatic carnival of non-evolutionary programmes in the wider behavioural sciences?

Conceptual and methodological pluralism remains the order of the day. Disciplines are simply too specialised and address too many different questions to be unified. Indeed, many philosophers of science believe unity of the sciences is not possible even in principle.

This exposes the fundamental weakness of proposing evolutionary psychology not only as a paradigm but as *the* paradigm of the behavioural sciences: it promises far more than it can possibly deliver, simply dismissing out of hand deeply established positions and issues. The majority of economists, sociologists, and psychologists hold theoretical beliefs that evolutionary psychologists do not consent to not because of ignorance of evolutionary theory or owing to artificial institutional barriers, but for reasons leading evolutionary psychologists barely acknowledge. In short, the championing of evolutionary psychology as the metatheory of the behavioural sciences is superficial; it has little persuasive force. It appears to be little more than marketing.

2.9 Policy and Popularisation Dangers

What about public policy? Has evolutionary psychology had much impact? Again, despite the promise the impact has been negligible. Policy makers consult economists, sociologists and various psychologists. They don't consult evolutionary psychologists.

But there is one area that evolutionary psychology has broken through with great fanfare. Although evolutionary psychology has yet to seriously impact policy it nevertheless has a regular impact on the popular imagination via mass media. Rarely a week will go by when not only tabloids, but also broadsheets, will sensationally report on the latest published research, to say nothing of such dissemination of such research on various popular websites, which probably has as great an outreach as the aforementioned.

Rarely does a social science disciple have such regular and reliable visibility in the mass media. Rarely does a social science discipline have a regular audience with an apparently insatiable appetite, a curiosity to read and to understand its research. Indeed, if any programme of research in the behavioural sciences has a claim to being the most popularised, then it's surely evolutionary psychology. It might not be the metatheory of the social sciences. But it has a quality of a metanarrative in the public sphere.

Indeed, Buller (2005: 3) begins his long and detailed critique of evolutionary psychology with precisely this observation.

> During almost every wait in the supermarket checkout line, I would find reference to the evolutionary psychology of human mating on the covers of women's and men's magazines. On Sundays, I would often find articles in the *New York Times Magazine* or the *Chicago*

> *Tribune Magazine* about the evolutionary psychology of mating, parent–child relationships, or status seeking.

Buller continues that such an outreach is not restricted to ink and paper but also to television. And not just specialised, supposedly high-brow documentary outlets, but also mainstream news outlets, special news reports examining the evolutionary psychology of sex differences and its implications. 'It appeared that evolutionary psychology was capturing the public consciousness and beginning to condition the human self-conception presented in popular culture' (*ibid.*)

And it is precisely this area, the understanding and popularisation of evolutionary psychology, its impact on popular culture, and its potential (if not actual) impact on policy, that has been the source of the most lively and animated contention, where debate in evolutionary psychology can slide from the polite to the unpleasant and the ugly. To understand why, consider the methodology. Consider reverse engineering a known trait or psychological or behavioural phenomenon. As presented, what's to stop an incorrectly identified trait or behavioural phenomenon—incorrect in the sense of being an artefact of flawed measurements or perhaps being merely an unsupported stereotype—from entering the adaptationist machinery?

Consider this: Do gentlemen prefer blondes over brunettes? Many people do think that men prefer blondes. Perhaps there's an adaptationist explanation for that trait? Ramachandran (1997) proposed precisely that. Ramachandran notes that the preference for blondes seems to be universal, being present not only in Western societies, but also in non-Western societies. So in virtue of its universality it could be an adaptation. How are we to explain blonde preference as an adaptation? People with blonde hair tend to have lighter skin and lighter skin makes skin infection more visible and so a preference for blonde hair would have been selectively advantageous.

Ramachandran (1997) was actually satire—not in the sense of the infamous Sokal hoax, of being a meaningless jumble of words published in a key journal, but in the sense of the hypothesis being crafted precisely to expose the thinness of adaptationist explanation. To some ears, the explanation can sound credible. And that's the point: stereotypes reworked into a superficial evolutionary spin can sound very persuasive. Perhaps men being more attracted to blondes over brunettes is relatively harmless, a leveraging of a common but not especially harmful stereotype. But what if the unsupported stereotype under consideration is more harmful? What if a corrosive stereotype was to undergo an adaptationist treatment? For example, what if one suggested that men tend to prefer white women to non-white women and that there is some evolutionary rationale for this?

It's possible for unsupported and even dangerous stereotypes to enter the adaptationist explanatory machinery as it has been described *so far*. And, unfortunately for evolutionary psychology's reputation, such a thing has been perceived to have occurred in a few well-known cases. Indeed, the reputation of evolutionary psychology has become toxic in some quarters owing to certain figures. We need not name any names.

Rightly or wrongly, in Western societies, scientific claims have a degree of prestige generally not accorded to non-scientific claims. Governments give pennies to poets but billions to scientists. Scientific claims are typically received with greater attention and less scepticism than claims from other sources.

This privileging of purported scientific explanations carries with it a risk of feedback loop or reflexivity between theory and social phenomenon, especially when those explanations are purportedly about human nature. There is a possibility that evolutionary psychology is simply reframing existing ideas in a *prima facie* scientific language, and that when this is broadcast back to the public via the media's apparent insatiable appetite for evolutionary psychology this will only strengthen those already conditioned attitudes yet further, which will again feed back into hypothesis generation and confirmation. Dupré (2003) highlights the fascinating example of Frank *et al.* (1993). As currently practised, economics is primarily concerned with investigating a range of self-interest models. Frank *et al.* (1993) found that such a focus on self-interest models makes those who study them less likely to co-operate—thereby supporting the very models under study.

All this shows that it's possible for incorrectly measured, incorrectly described, merely stereotypical and even dangerously stereotypical behavioural phenomena to enter through the rickety adaptationist door and to be explained in adaptationist terms. Through the rickety door of adaptationism and the eagerness to convey ideas to the public there is a danger that contingent beliefs will not only be promoted as being something 'anchored' in our 'nature', but also re-energised and reinforced in our psychological conditioning. It shows that it's even possible for battered, dying stereotypes to be resurrected with new life and vigour, to have to be fought again by those who have taken it upon themselves to fight such stereotypes—hence, perhaps, why evolutionary psychology seems to be especially challenged by those who actively and energetically identify themselves as progressives.

One can now see why evolutionary psychology has been attacked on the basis of its attempt to transport itself into public policy, despite its explanatory thinness, owing to perceived dangers that can result from its methodology. Dupré (2001, 2003, 2012) has especially pressed the point that the 'misguided project' of evolutionary psychology is not only explanatory weak, but also politically dangerous. Rather than be a mirror of the past, it can be a mirror of our prejudices—something science should be challenging, not crystallising.

Of course, this is not to say that explanations should be rejected on the basis of having negatively perceived implications. But it is to say that if the adaptationist doorway can let mistaken or dangerous perceptions to go through, if adaptationist explanations are advertised as being more solid than they actually are, and if they are frequently conveyed to public, who often receive them uncritically owing to their scientific appearance, then this is going to generate dramatically more resistance and heat than is normally to be expected from an academic debate.

2.10 Conclusion

We began by identifying and correcting five common areas of misunderstanding. First, it's a mistake to suppose that the levels of selection debate problematises evolutionary psychology. One could pursue evolutionary psychology along neo-group selection lines without difficulty. Furthermore, kin selection and neo-group selection are actually different ways of looking at the same thing, and there are good reasons to prefer kin selection over neo-group selection.

The second issue we looked at was the EEA. The EEA is part and parcel of the concept of adaptation. It should not be controversial. It only becomes controversial when the possibility of mismatch confused with the ubiquity of mismatch. This conflation has been allowed to waste a lot of time and has bred a distorted picture of evolutionary psychology. Furthermore, evidence of recent selection for simple physiological traits does not destabilise the argument that there's been insufficient time for selection for complex psychological traits; not a single recently evolved complex psychological trait has been proposed; even if selection could select for complex psychological traits in a short period of time, the instability of recent social environments problematises their selection; even if new complex traits have recently been selected for, this doesn't undermine the search for complex traits with longer pedigrees.

Third, behavioural flexibility. If psychological adaptations are species-typical, how do evolutionary psychology hypotheses account for variability? Psychological adaptations are not blind instincts. They generate variable behaviour as a result of their being selected to respond adaptively to variable ancestral conditions. Indeed, as we'll see in the next chapter, evolutionary psychologists do not merely pay lip service to variability of behaviour—much of their research focuses on the specifics of how people respond contingently to environmental conditions.

Fourth, developmental complexity. Evolutionary psychology is an arena where the old antagonisms of genes versus environments genuinely no longer pertain. It is quite possible to endorse causal holism cheerfully enough while also postulating developmentally robust adaptations. Indeed, as we shall see in the next chapter, development is important for evolutionary psychology thinking in another way: early experience can developmentally calibrate psychological adaptations by, for example, calibrating the thresholds for evoking certain responses. Thus, we have another dimension of variability: you might be developmentally calibrated one way, and I might be developmentally calibrated another way.

Finally, when some prominent evolutionary psychologists use the term 'module'—I say prominent evolutionary psychologists, for the term 'module' is not frequently employed by practitioners on the ground, nor does it need to be employed—they simply mean functional specificity, not Fodor's use of the term to mean informational encapsulation.

Nevertheless, it is not possible to simply reduce the case against evolutionary psychology to a case of conceptual misunderstanding. The shift from evolutionary psychology as a research programme to that of a paradigm begins with the claim that selection would likely have favoured a massively modular architecture. Stronger versions claim that selection would have selected against domain-general mechanisms or domain-general intelligence.

But what should have been obvious at the outset is that one can't simply read off from empirical adaptationism the likelihood of a particular cognitive architecture having evolved. The inference just isn't one directional: a variety of different cognitive architectures are consistent with empirical adaptationism. Empirical adaptationism simply doesn't make more probable a massively modular architecture in the way leading evolutionary psychologists suppose. And this foreshadows a deeper, methodological challenge: concerning the degree of indeterminacy inherent in the method of reasoning from adaptive problem to adaptive solution and vice versa. And in light of the standards to be expected of complete adaptationist explanations, evolutionary psychology explanations have deplorable deficiencies. Indeed, this should not be surprising: reasoning between adaptive problem and adaptive solutions alone cannot possibly provide the kinds of evidence needed to deepen and vindicate such explanations. Accordingly sceptics invite us to see these explanations for what they believe them to be: just so stories, speculation. Nothing to rationally command assent to.

Leading evolutionary psychologists believe in the need to unify the behavioural sciences and they believe evolutionary psychology can do this. But both the need to unify and the ability of evolutionary psychology to do this are opaque. Unifying the behavioural sciences, assuming that this is even needed, is a herculean endeavour; it can't be done with a few thoughts tagged in theoretical manifestos or edited volumes. Furthermore, sceptics invite us to consider that the shaky method not only produces indeterminate and thin explanations, but could also be politically and socially dangerous—and so must be exposed and resisted.

All of this forms a powerful impression that, no matter its purported benefits, evolutionary psychology is discredited, something one can legitimately dismiss wholesale. As we shall explore in the subsequent chapters, I do not believe that such objections, as powerful as they are, mark the end of evolutionary psychology as a research programme. But they do mark the unsustainability of evolutionary psychology as a paradigm.

3
Evolutionary Psychology and Novel Predictions

3.0 Introduction

So far, the case for evolutionary psychology is not looking promising. As a paradigm, evolutionary psychology is problematic, contentious and antagonising, being fiercely scrutinised and challenged. Its explanations are exposed as embarrassingly naked in historical and scientific detail—the kind of detail required to vindicate them. If that is all there is to evolutionary psychology, one should think the case against it is decisive. Indeed, many have come to precisely that conclusion. But there is another dimension to evolutionary psychology, a dimension relatively overlooked, and indeed often not even recognised, in the literature: evolutionary psychology as a heuristic, a set of tools for discovering new features of human psychology and behavioural phenomena.

All the way back in 1987 Cosmides and Tooby were writing that evolutionary theory can be used 'as a heuristic guide for the discovery of innate psychological mechanisms' (1987: 279), and later wrote that this is the 'essence' of the programme (Ermer *et al.*, 2007: 154). Symons proposed that adaptationist thinking in psychology 'can help guide research', 'inspire new questions' and 'call attention to aspects of human psychology that are normally too mundane or uniform to notice' (1989: 143). Barkow says 'Evolved mechanisms are useful heuristic devices' (2006: 21). Buss frequently speaks of 'the heuristic value of evolutionary hypotheses' (2005b: 252) and that these heuristics are 'powerful' (2005c: 1). Daly and Wilson concur, also speaking of the 'heuristic assistance of Darwinian insights' (1996: 22) and that these heuristics are 'strong' (2005a: 408).

And yet reading the standard accounts of evolutionary psychology that present it as a paradigm you might miss this. The heuristic function was mentioned in the previous two chapters but the other tenets overshadowed it. And indeed that's precisely what seems to have happened in the debate: the eye-catching spectrum of tenets has made the heuristic component relatively invisible. This is the weak point of sceptical treatments of evolutionary psychology: most ignore this dimension and the few who do recognise it hand wave it away. It's time to address this oversight.

The basic thought motivating this chapter is as follows. Suppose natural selection has shaped our psychological mechanisms. If that is so, then adopting an adaptationist perspective in psychology should highlight features of our psychology that we have hitherto been unaware of, features we might otherwise miss entirely. Indeed, we might even be able to discover hitherto entirely unknown psychological mechanisms. If evolutionary psychology is capable of making such novel predictions and has successfully done so, the situation is dramatically changed: finally we will have a solid foundation, an unchallengeable basis, for evolutionary psychology. If—for it's by no means an *a priori* guarantee that adaptationist thinking can furnish such predictions. The case that it can and does must be fought for and won.

This chapter seeks to win evolutionary psychology such a case. In 3.1, I set out what heuristics are and how we can judge whether a research programme or effort has heuristic value. In 3.2, I highlight the centrality of the good design or engineering heuristic, the engineering demand for form–function fit, in evolutionary psychology and I discuss in more detail the two broad research strategies evolutionary psychologists use to generate novel predictions. In 3.3, I show how these two research strategies work in tandem—a crucial point largely absent in the literature. Evolutionary psychology hypotheses might start off as 'simple' but they can progressively become more complex, progressively mirroring the adaptations they're targeting. Hypothesis generation in evolutionary psychology is not a one-shot event but a continuous, bootstrapping process. Sections 3.4 and 3.5 explicate further heuristics available to evolutionary psychology for identifying adaptive problems and adaptive solutions respectively. Section 3.6 looks in more detail at exactly the kind of experiments evolutionary psychologists can and do perform to confirm or disconfirm their novel predictions. After all this, we will be in a position to fully address the methodological objections raised in the previous chapter, and we shall do this in 3.7. First, evolutionary arm races do not problematise either the concept of an adaptive problem nor adaptive reasoning between problem and solution—rather evolutionary arm races mean that many of our psychological adaptations are likely to be richly calibrated, and adaptationist methodology is entirely capable of capturing such dynamics. Second, the grain problem is an illustration of something being an explanatory sin but a heuristic virtue: granularity can destabilise incomplete explanations but prove fruitful to discovery. Third, the design heuristics dramatically constrain research space while still allowing creativity within the constrained case. Fourth, the cases used to argue against evolutionary psychology being heuristic in practice are old cases, from the beginning of the programme, precisely when one would expect activity to be primarily one of accommodation. The research programme is over two decades old and has produced a succession of successful novel predictions, to an extent significantly greater than is generally appreciated.

The goal is clear: to establish that evolutionary psychology is heuristic in both principle and in practice. As one might already begin to imagine, this could have dramatic ramifications. We shall explore those in the next chapter.

3.1 Heuristics in Science

Sometimes we face a number of options so vast that we need a way of cutting them down—ideally to good options. Here is a simple thought experiment, based on a scene from a well-known movie. Suppose you're an adventurer seeking the Holy Grail—the legendary chalice from the Last Supper. You enter a chamber and you behold hundreds of grails, an assortment of different types: silver grails, gold grails, ruby encrusted grails, small grails, large grails and so on. A guardian tells you the following: drink from the Holy Grail and you shall surely become immortal; drink from a false grail and you shall surely die. Circumstances are such that you are forced to choose a grail.

Out of hundreds of possible choices, which grail do you choose? Trial and error is simply not possible. So how do you choose? What kind of consideration would deliver the best bet to make? Pick the most dazzling grail? The largest? The smallest?

How about this: perhaps a good basis to make such a perilous decision would be to use knowledge of history or legend. You think about the origins of the grail: Who was the grail for? In what circumstances? You might conclude that a humble man would have a humble cup. You then scan the hundreds of chalices and see an unassuming cup. Nothing extraordinary, just an ordinary cup common in the Middle East many centuries ago. You make your choice, your bet. You select this grail. You live.

This simple thought experiment vividly illustrates the function of heuristics, as well suggesting their key features. When faced with a large number of options, our adventurer used a reasonable strategy to cut them down dramatically, to identify a plausible option—but without certitude. The word 'heuristic' simply means to find or to discover. In the context of science, heuristics are recipes, strategies or methods for generating novel hypotheses in a given domain. They work by reducing research space, focusing research effort along constrained, plausible, but uncertain trajectories. By fleshing out these defining properties or ingredients—hypothesis building, predictive novelty, constraint and risk—we can establish a benchmark for adjudicating whether evolutionary psychology has heuristic value.

First, heuristics guide and organise research—specifically, they guide and govern hypothesis generation. The conceptual framework of a heuristic research programme will specify the objective of the research effort, what its targets are, such as detecting and finding certain patterns or phenomena within a given domain. The methodological framework will be crafted according to the particularity of the objective—that is, it will present a recipe or set of strategies for achieving the research goal.

Such recipes or strategies, the heuristics, will specify how to initiate a relevant research trajectory—what constitutes a legitimate starting point, what things to consider and what things not to consider. A series of methodological steps or stages will be presented, which, when followed, generate a research hypothesis or solution. The steps can take a positive form (consult *X* information) or a negative form (be wary of *Y* information). Some steps will be obligatory—certain questions must be asked; certain information must be consulted and taken into account. Other steps will have a degree of optionality, being more like useful hints than requirements (e.g. if *X* and *Y* are the case, consider *Z*). Optional steps usually come into play when refining hypotheses. The end goal—the testable hypothesis—must satisfy some specified criteria.

A research programme's heuristics will often be presented in theoretical manifestos and introductions to edited volumes, often by the leading practitioners of the programme, the trailblazers, and often with one or more flagship examples of successful hypotheses delivered by such means. Naturally, therefore, these are the first port of call for those wishing to understand a research programme's heuristic resources. But here caution must be exercised: one mustn't fall into the mistake of thinking that these are necessarily complete or comprehensive accounts. They might be, and perhaps often are, only partial codifications. Some heuristics might be implicit and only occasionally articulated. Furthermore, the stages of hypothesis generation might be more sophisticated, more dynamic, than is gleaned from reading standard, introductory accounts. The philosopher who wishes to adjudicate on the heuristic value of a given programme should therefore be wary of consulting such accounts in isolation and then make play with perceived shortcomings—rather, she should examine such accounts in tandem with best practice, and submit a more comprehensive methodological account if existing accounts are incomplete.

Second, heuristics help researchers generate novel predictions. Heuristics are associated with the exploration of hitherto unexplored research paths. They allow researchers to see a domain or subdomain in novel ways, to investigate possible patterns that those without the heuristics are blind to and to articulate these unique insights to the point of testability. So one of the hallmarks of heuristics is that they generate genuinely *novel* hypotheses: they generate novel predictions about known phenomena or novel predictions about hitherto unknown phenomena. Novel expectations violate our expectations. The more novel these predictions are, the more surprising and counterintuitive, the better. As with all concepts, the concept of novelty admits a degree of ambiguity, and different versions can be cashed out. Musgrave (1974) distinguished three versions of novel predictions: temporal (a phenomenon is novel for a hypothesis if it was unknown at the time of the hypothesis's formulation), 'heuristic' (a phenomenon is novel for a hypothesis if the hypothesis was not constructed specifically to accommodate the phenomenon) and theoretical (a phenomenon is novel for a hypothesis if it's not predicted by any existing rival hypothesis). For the purposes of our discussion, we shall understand novel predictions in the so-called temporal sense.

If one seeks to establish a research programme's heuristic credentials, one must establish that in principle the strategies and methods can generate novel predictions, and that in practice it has indeed generated genuinely novel predictions. Likewise, if one seeks to deny that a research programme has heuristic value, one option is to scrutinise and challenge novelty claims. One could challenge whether the methods really have yielded novel predictions. Perhaps behind the fanfare and fireworks, a research effort is simply accommodating what we already know within its conceptual framework.

Third, heuristics drastically cut down (often quite vast) research space, reducing the search space for a given research objective to something plausible and manageable. Suppose our research objective is to discover new facts about extant psychological traits. The search space, the space of all possible hypotheses, will be vast—too vast to execute an exhaustive trial and error search across all possible trajectories and avenues. What we need, therefore, is something to constrain this search space, identifying plausible or promising trajectories, leaving implausible and less promising trajectories unexplored. Heuristics determine what to look for—what constitutes a legitimate starting point, what methodological steps must be taken, what criteria a proposed hypothesis or solution must satisfy—and so perform this function, restricting precious research effort to a relatively small portion of the possible research space. So to establish whether a research programme has heuristic value, we need to establish that its strategies and methods adequately constrain research space.

That heuristics constrain research does not mean there is no room for creativity. Heuristics are constrained but creative ways of exploration. They dramatically limit the search area but there is still a degree of creative exploration within the constrained search area. Once a heuristic is adopted, certain pathways will standout—but there will still be space within those pathways to explore.

Therefore, we should not read the constraint requirement as a requirement that heuristic strategies and methods produce only one hypothesis for a given target. It's possible, but by no means necessarily typical, for two researchers following the same set of procedures to generate two different hypotheses for a given target. Heuristics consist of steps and procedures and solution criteria and, while they constrain hypotheses, creative differences of interpretation will occasionally generate a handful of hypothesis for a given target instead of just one. The point, however, is that heuristics allow only a relatively limited number of hypotheses to be generated for a given target, whether one or occasionally a small number—a small number at most, suitable for experimentation, not a paralysing, runaway list of hypotheses. Furthermore, even when a competition between several possible hypotheses is resolved in favour of one, strategies and methods might suggest or recommend several options for refining and developing the successful hypothesis, which might necessitate a limited amount of tinkering and trial and error. Again such permutations will not be too great, being within a suitable range to investigate.

So even while heuristics significantly reduce the size of the research space, it's likely that researchers will deploy degrees of creativity and degrees of trial and error within the constrained space. In sharp contrast, strategies and methods that can be used to generate a runaway list of hypotheses for a given target, that generate activity seemingly indistinguishable from blind trial and error, do not have heuristic value. So in the example of seeking to discover new facts about a given psychological trait, strategies and methods that allow an avalanche of competing hypotheses to be generated about the trait cannot be said to be heuristic.

Fourth, heuristics are risky: they constrain research to plausible and promising trajectories without guaranteeing such exploration will be empirically successful. 'The search space is vast, so the method of search must be "heuristic"—the branching tree of all possible moves has to be ruthlessly pruned by semi-intelligent, myopic demons, leading to a risky, chance-ridden exploration of a tiny subportion of the whole space' (Dennett, 1995: 209). Heuristics will give us a range of research bets to make—reasonable bets, worthwhile bets, but bets all the same, trajectories involving risks of failure, of being dead ends. To borrow a classic Popperian phase, heuristics 'stick their necks out'.

So we can expect heuristic strategies to lead occasionally to predictive failure. Inevitably there will be disappointment. Nevertheless, they are not expected to systematically fail. Some, perhaps even many, predictive bets won't pay off—but, crucially, some predictive bets will and it's those successful bets, those bets that become lasting wins, that justify the enterprise and give it value. This is why heuristics can win a research programme scientific legitimacy: by charting risky, unexplored but plausible trajectories, by making successful novel predictions, research efforts that employ heuristics make genuine and permanent editions to knowledge, establishing new facts that would otherwise be missed in the sea of possible hypotheses.

When introducing the notion of heuristics in science, it will not have escaped your notice that our adventurer was using historical knowledge to cut down dramatically his options, to identify the most plausible option. A bet, not a guarantee. He was using historical considerations as a heuristic. And it is precisely this kind of strategy that is adopted and championed by evolutionary psychology.

3.2 Heuristics in Evolutionary Psychology

So far, I have been allowing the description of evolutionary psychology methodology as it often figures in the literature, especially in the sceptical literature, a description that doesn't do full justice to its power, potential and dynamic character. It's now time to provide a more refined, more detailed account.

The underlying logic of evolutionary psychology methodology is this: *if* psychological trait T is a psychological adaptation for X, then psychological trait *T* should have configuration *C* and so we should find phenomenon *P*. That is, if *T* is an adaptation for *X*, we should see signatures of this in the adaptation's design, in the kinds of inputs it receives, in the kinds of output it produces and so on, and this can locate behavioural patterns we've never observed before. Campbell (2006: 91) nicely cashes out this logic concretely and succinctly as follows:

> An adaptation (such as fever) designed to produce a particular beneficial outcome (parasite destruction) should occur when the body detects pathogens and not at other times ... If we have conserved a herding response to danger, we should prefer to be with others when we face threatening uncertainty. If guilt is an emotion signaling unpaid obligation, we can make clear predictions about the likelihood of altruism when it is aroused.

Recall that in Chapter 1 we cited Buss's (1995) definition of a psychological adaptation (what he terms 'evolved psychological mechanisms'): a psychological adaptation is a set of processes that (1) exists, because it solved a specific and recurrent adaptive problem; (2) receives only certain information inputs and converts that information into output through a procedure that adjusts physiological activity, provides information to other psychological mechanisms or produces overt action.

We can say that (1) pertains to the function of a psychological adaptation and (2) pertains to the form of a psychological adaptation. Evolutionary psychology methodology leverages the possibilities generated by this definition. In generating hypotheses, evolutionary psychologists heuristically reason between function and form of proposed psychological adaptations. Or, to put it another way, they heuristically reason between ultimate questions about the selected trait under investigation (why the trait exists, what it was selected for) and proximate questions about the selected trait (how it works: what information it processes, what outputs it produces).

In Chapter 1, we briefly saw that two research strategies are available: what we initially called 'reverse adaptationist reasoning' and 'forward adaptationist reasoning'. A more refined way of framing these strategies is in terms of function, engineering and design. Hence, we can speak of 'reverse engineering', where we reason from form back to function, from the present to the past. Here we begin with an extant psychological trait and then consider what kind of adaptive problem it's well designed for solving. And we can call the second strategy 'forward engineering', where we reason from function towards form, from the past to the present. Here we begin by postulating an adaptive problem and then reason what kind of psychological trait would constitute a well-designed solution to it. Notice that this second strategy, of reflecting on adaptive problems and scenarios, holds out the possibility of discovering new mechanisms. This possibility afforded by forward engineering achieves its sharpest expression in a passage in Ermer *et al.* (2007: 154):

> Evolutionarily rigorous theories of adaptive function specify what problems our cognitive mechanisms were designed by evolution to solve, thereby supplying critical information about what their design features are likely to be—information that can guide researchers to discover previously unknown mechanisms in the mind. That is the essence of the adaptationist program.

On the strength of these considerations, I postulate the following two heuristic functions of evolutionary psychology:

- H1 Hypothesising unknown design features of extant psychological mechanisms, leading to novel predictions of psychological phenomena
- H2 Hypothesising unknown psychological mechanisms, leading to novel predictions of psychological phenomena

Evolutionary psychology is about being sensitive to the design features of psychological traits and being alive to the possibility of hitherto unknown design features of known psychological mechanisms and even to the possibility of hitherto unknown psychological mechanisms awaiting discovering. To operationalise the heuristics of evolutionary psychology is to be able to see behavioural phenomena in this unique and predictively fertile way. This, I contend, is the primary function and value of evolutionary psychology.

And this is precisely why evolutionary psychologists proceed to design and carry out psychological tests in first place. If, as their critics frequently suggest, evolutionary psychologists simply aim at explanation, at simply explaining observed phenomena in adaptationist terminology, then there would be no need for them to design and perform psychological tests (for such tests would simply confirm the existence of the phenomenon to be explained): they could simply come up with a selection story, append any studies already known to be consistent with the explanation as support and leave it at that.

But as evolutionary psychology is actually investigating whether psychological traits have features to be expected if they are well-designed solutions to adaptive problems, this necessitates the design and execution of experiments that aim to verify such design features. The successful corroboration of such novel predictions will expand the explanandum of the purported behavioural phenomenon or psychological trait underpinning it. So even if the target of an evolutionary hypothesis happens to be an already known behavioural phenomenon or, more precisely, the psychological mechanism underwriting it, evolutionary psychologists are still trying to discover previously unknown features of that mechanism, features invisible or dim in the light of ordinary thinking, features visible in light of form–function thinking.

So evolutionary psychology hypotheses are motivated by, and are best understood on the basis of, heuristic ambitions. One could choose to be entirely non-committal or agnostic about the explanatory status of these hypotheses. Or, as seems more common, one could view them as tentative, preliminary explanations, which triggers the legitimate explanatory and evidential demands that Richardson (2007) and others have identified and tabled. We'll look at the challenge of adaptationist explanation in next chapter. For now, our focus will be exclusively on evolutionary psychology as a heuristic.

Notice that the key focus of the evolutionary psychologist, the key questions she asks, is not at the level of selective fitness, not on what kind of trait in general would have been selectively advantageous. The focus and set of questions goes deeper: What would have been a well-designed solution to the adaptive problem? Design and form–function fit, not fitness, is at the forefront of evolutionary psychology hypothesising. Of course, fitness underwrites design—if a psychological trait solves an adaptive problem, then it will be selected for in virtue of increasing fitness—but fitness considerations are at the background, not forefront, of evolutionary psychology heuristic thinking. And as we shall see later, this is crucial, for while fitness thinking can be elastic, possibly generating a runway list hypotheses, form–function thinking dramatically reduces the number of hypotheses that can be generated.

An example of H1 is Curtis *et al.* (2004), which hypothesised that disgust is an adaption for avoiding pathogens. If the function of disgust is to act as a pathogen-avoidance system then a number of novel predictions can be generated. First, we can expect the feeling of disgust to be dramatically stronger when one encounters cues with disease connotations than when one encounters similar cues but lacking any disease connotation (e.g. seeing a picture of bodily fluids versus seeing a picture of a blue chemical dye—similar cues but only the former has any disease connotation). Second, we can expect disgust to operate in a similar way across cultures. Third, we can expect a sex difference at the level of design: as women protect not only themselves, but also offspring, disgust should be more pronounced in women. Fourth, we can expect disgust should be less pronounced as an individual's reproductive potential declines. Fifth, as strangers carry a higher risk of carrying novel pathogens, disgust should be more strongly evoked by contact with strangers than close relatives. Curtis *et al.* (2004) tested these predictions using data provided by almost 40,000 participants using an international website survey employing visual stimuli, where individuals were asked to grade the disgustingness of seven pairs of visual stimuli. For each pair of visually similar stimuli, one had disease connotations and the other didn't. Cross-culturally, the disease-relevant stimuli were found to be significantly more disgusting than their disease irrelevant counterparts. The authors concluded that the results are consistent with disgust being an adaptation for disease avoidance.

An example of H2 is the previously unknown cue detection of the ovulatory cycle. In contrast to many other mammals, including our closest cousin, the chimpanzee, human ovulation is concealed. The ovulation of chimpanzees is advertised by a sizable swelling of the genitals. Evolutionary psychologists hypothesise (1) the possibility of subtle cues indicating stages of the ovulatory cycle, and (2) the possibility that natural selection selected for adaptations dedicated to detecting those cues. Recent research is progressively vindicating both (1) and (2).

With regard to (1), emerging research is identifying discernible cues of fertility in women's movements, scents and voices. For example, the voice attractiveness of women varies across the ovulatory cycle (Pipitone and Gallup, 2008). With regard to (2), evidence suggests that women's male partners adaptively shift their behaviour in response to cues of approaching ovulation. Evidence has already shown that men subconsciously judge where a woman is in her ovulatory cycle. For example, lap dancers make significantly more money in tips when they're ovulating compared with when they're menstruating (Miller *et al.*, 2007).

This is a unique prediction. To some, it might sound a little far-fetched, somewhat difficult to believe. It's easy to be dismissive of such ideas and findings. This is an important theme, one shedding light on the importance of evolutionary psychology in the evolutionary and behavioural sciences, and shedding light on a key source of scepticism of the programme. I'll return to this theme in the final chapter.

No matter whether we use reverse or forward engineering, we need to consider a number of selectionist and engineering questions. What kind of selection pressures obtained in past ancestral environments? What kind of recurrent adaptive problems and opportunities did they create? When considering what would serve as a well-designed solution to such a problem, what kind of information, what kind of cues, would potentially have been available in past ancestral environments for solving such a problem? Are these cues reliable indicators? What kind of computations would a well-designed solution to the adaptive problem make? What would it do with the inputs? More precisely, what feelings, physiological activities, thoughts and actions would it produce? What kind of developmental and contextual calibrations would be required, if any, to solve adequately the adaptive problem?

So adopting an engineering perspective, asking what a well-designed solution to a given adaptive problem would look like, thinking about the fit between adaptive problem and adaptive solution, is the cornerstone of adaptationist hypothesising in psychology. Cosmides and Tooby (1997a: 15) write:

> engineers figure out what problems they want to solve, and then design machines that are capable of solving these problems in an efficient manner. Evolutionary biologists figure out what adaptive problems a given species encountered during its evolutionary history, and then ask themselves, 'What would a machine capable of solving these problems well under ancestral conditions look like?'

Potentially there is much to be gained by hypothesising a close fit between adaptive problems and adaptive psychological solutions. It's commonplace that often the most difficult thing to see is right in front of your nose. Hidden in plain view. We need theories and hypotheses to guide us into identifying patterns. Evolutionary psychology can guide researchers into predicting new behavioural patterns, new demographic patterns, new facts—facts that do not readily lend themselves to discovery by non-adaptationist perspectives. The more novel these predictions are, the more surprising and counter-intuitive, the better.

Of course, many psychological traits are not adaptations. Evolutionary psychologists don't take just any random psychological trait and start hypothesising with it. The heuristics constrain what to look for. Evolutionary psychologists look out for putative or candidate adaptations, where there is evidence, or possibility of gaining evidence, of either (1) special design, (2) universality, (3), ease, speed and reliability of development or learning, or (4) maladaptiveness. Evidence of any of these indicates that a trait *could* be an adaptation, not that it *must* be an adaptation, and thus warrants further adaptationist analysis.

Special or adaptive design is recognised when a trait exhibits complexity, selectivity, efficiency and specificity in its outcomes (Williams, 1966). What is sometimes called 'Dawkins' gambit' holds that complexity of design is a clear and unambiguous sign of selection: complex traits or behaviour patterns cannot evolve by chance or drift but must have been selected for and shaped by natural selection. However, as Richardson (2007) points out, some biologists challenge this strong inference. We need not get drawn into this debate. Although the gambit is present in prominent evolutionary psychology literature, such a strong inference is not strictly required: we can simply state that special design of a trait signals the *possibility* that selection has acted on the trait.

Adaptations are species-typical (universal) at the level of design. Hence, if a trait is found universally—across cultures, spatially and temporally—this is also suggestive that the trait could be an adaptation. Note, however, that a trait being cross-cultural does not entail it will be expressed the same across cultures; different environmental and social conditions can supply different inputs that result in the trait generating different behavioural outputs. The point is that the variability of expression based on local environmental (ecological and social) and developmental conditions should be predictable in virtue of species-typical design.

Traits that develop especially quickly, traits that are learned especially easily, traits that develop especially reliably can also be considered as candidate adaptations. Examples include the capacity to develop language at a young age, the ability from a young age to recognise other individuals by their faces, and inferring the feelings and intensions of others from facial expressions, and a biased propensity to fear snakes.

Furthermore, if a trait is deleterious in today's environment, this could indicate that it's a well-designed adaptive solution to a past adaptive problem but simply maladaptive in today's environment. As we saw in Chapter 2, certain traits are strikingly maladaptive—famously, our preference for fatty and sugary foods.

It's important to note that an evolutionary psychology analysis of a psychological phenomenon does not require the phenomenon to be directly underwritten by a dedicated adaptation. Evolutionary psychology can apply to phenomena that emerge as by-products of other psychological adaptations—and surely this could be a large class of phenomena.

In light of the distinction between adaptations and by-products, mentioned in Chapter 1, evolutionary psychology has a further use:

H3 Hypothesising traits as by-products of psychological mechanisms, leading to novel predictions of psychological phenomena

A good example of H3 is evolutionary psychology work on the psychology race, as undertaken by Kurzban *et al.* (2001) and Cosmides *et al.* (2003).

Previous research had established that people automatically encode the race of individuals they encounter. Race exists in our minds. The problem with this, observe Kurzban *et al.* (2001), is that there are no objective patterns in the world that could explain why racial categories are so salient, so obvious to adults.

The puzzle deepens when one realises that in the environment of evolutionary adaptedness (EEA), there were no adaptive reasons to encode race automatically. It's plausible that selection would have favoured automatic encoding for an individual's sex and age. However, it's unlikely anyone would have encountered other people so distinct as to be perceived as a different 'race'. Ancestral hunter–gatherers had a limited travel range, having residential moves no greater than 40 miles. 'If individuals typically would not have encountered members of other races, then there could have been no selection for cognitive adaptations designed to preferentially encode such a dimension, much less encode it in an automatic and mandatory fashion' (*ibid.*, 2011: 15387).

The psychology of race must therefore reflect some other psychology. The adaptationist question then becomes: What was the adaptive problem faced by our ancestors that nevertheless apparently makes race so salient? Kurzban *et al.* (2001) hypothesised that it was the adaptive problem of making coalitions. Hunter–gatherers lived in bands. Between bands and within bands conflict can arise and coalitions and alliances can form. Hence, there was an indispensible pressure and need to track the ever-shifting alliances in the social environment—to be alert to fluctuating coalitional cues.

Hence, there was an adaptive problem for detecting coalitions and alliances. Therefore, Kurzban *et al.* (2001) hypothesised that we have a psychological adaptation for tracking coalitions and alliances. Function–form reasoning led to the hypothesis that the proposed coalition detection mechanism will be sensitive to a variety of cues suggesting group membership:

> Computational machinery that is well-designed for detecting coalitions and alliances in the ancestral world should be sensitive to two factors: (1) patterns of coordinated action, cooperation, and competition, and (2) cues that predict—whether purposefully or incidentally—each individual's political allegiances (Cosmides *et al.*, 2003: 177).

In today's world, consisting of racially diverse individuals living in close proximity to one another, race-related physical features may incorrectly be perceived as cues for coalitional alliances, hence resulting in the automatic encoding of race.

If there is such a psychological adaptation, a number of consequences follow. More precisely, Kurzban *et al.* (2001) derived six novel predictions, including the prediction that manipulating coalitional variables should introduce variability in encoding. In other words, people shouldn't encode the race of other people when it is not indicative of group alliance. This was tested by experimental manipulation in which race and alliance were decoupled. The result was dramatic: within just a few minutes the tendency to categorise by race faded.

> Despite a lifetime's experience of race as a predictor of social alliance, less than 4 min[utes] of exposure to an alternate social world was enough to deflate the tendency to categorize by race. These results suggest that racism may be a volatile and eradicable construct that persists only so long as it is actively maintained through being linked to parallel systems of social alliance (Kurzban *et al.*, 2001: 15387).

So, rather unexpectedly, the psychology of race can be understood as a by-product of a psychological adaptation for something else. This led to a new discovery: the automatic encoding of race can be eliminated by modifying contextual circumstances. Hence, according to Kurzban *et al.* (2001), the tendency to encode race automatically is not to be thought of as an adaptation. It is a by-product of a psychological adaptation evolved for other purposes—our psychological adaptation for detecting coalitional alliances. It is not inevitable.

But we can do more than just hypothesise a phenomenon as being a product or by-product of a singular psychological adaptation. We can also hypothesise a phenomenon as being a by-product of multiple psychological adaptations. Evolutionary psychology work on religion is an excellent example of this principle. The consensus view in evolutionary psychology is that religious belief and behaviour are not adaptive traits but rather by-products of several psychological adaptations.

Religion, a belief in supernatural agents and the practising of rituals, exhibits certain features that attract the attention of evolutionary psychologists. First, religion is a cross-cultural phenomenon. No matter what century, no matter what type of society we examine, religious activity seems widespread. Atheism is relatively rare, even in the West. Second, religion is robust. Despite religion being irrational or a-rational in the sense of lacking evidential or rational justification, religion persists, even in places where the general intellectual Zeitgeist has moved against it. Religion even persists in states that actively persecute against it. Early Christianity, for example, persisted and flourished within the Roman Empire despite Rome's severe persecution of it. It also survived Soviet persecutions in the 1920s and 1930s.

Beliefs and practices that are pervasive and persistent, and which lack an exclusively rational basis in terms of evidential warrant, might very well be drawing on, or be produced by, evolved psychological processes. Although a given trait being pervasive and persistent does not mean that it must have an adaptationist underpinning, either directly or indirectly, it does, nevertheless, represent a good reason for pursuing an evolutionary psychology investigation into the phenomenon. If it does indeed turn out to be the case that religion's vitality is drawing on evolved mechanisms, then an evolutionary perspective brought to bear on religion should offer fresh insights.

Despite exhibiting features that suggest a deep psychological undercurrent, religion also exhibits other characteristics that render it somewhat problematic from an evolutionary perspective. Most strikingly, religious practices appear to have high fitness costs. As Atran (2002: 4) succinctly puts it:

> From an evolutionary standpoint, the reasons religion shouldn't exist are patent: religion is materially expensive and unrelentingly counterfactual and even counterintuitive. Religious practice is costly in terms of material sacrifice (at least one's prayer time), emotional expenditure (inciting fears and hopes), and cognitive effort (maintaining both factual and counterintuitive networks of beliefs).

These costs must have been even more pronounced during our prehistory. The task facing the evolutionary psychologist, then, is to explain how religion could have arisen and persisted despite its costs.

Atran (2002), Boyer (2002), Barrett (2004) and others account for religion as the by-product of a convergence of several evolved cognitive systems interacting with the world. This evolutionary psychology approach to religion recognises that religion has fitness costs. However, these fitness costs are outweighed by the gains in fitness brought about by the cognitive systems that give rise to religion. Thus, although religion is not an adaptation, the cognitive systems that converge to give rise to religion as a by-product are adaptations. This approach holds that given how our brains have evolved, religion is almost inevitable: 'We do not have the cultural concepts we have because they make sense or are useful, but because the way our brains are put together makes it very difficult not to build them' (Boyer, 2002: 187)

What cognitive mechanisms are hypothesised as being responsible for giving rise to religion as a by-product? One of the most important mechanisms is our predisposition to perceive agency. Guthrie (1993) brought attention to a perceptual bias in humans: humans have a tendency to interpret ambiguous stimuli in terms of agency. Thus, we occasionally see, for example, faces in configurations of clouds, raindrops and even on Martian mountains. We hear a noise or rustle in the bushes and are immediately gripped by the expectation of an unknown agent nearby. Guthrie suggests that this perceptual bias is adaptive. Barrett (2004) called the cognitive mechanism producing this perceptual bias the 'hyperactive agency detection device' (HADD). The hypothesised HADD is understood to detect agency when an object is perceived to be contravening assumptions concerning the movement on inanimate objects (such as moving without another body moving against it). Having this perceptual bias to see agency obviously carries with it great increases in fitness. False positives carry little cost; failing to detect a predator, in contrast, carries a heavy price. Error management theory states that one should expect the evolution of biases that minimise costs of errors in judgement, even if such biases produce more errors. More precisely, error management theory predicts that, when decisions are made under uncertainty, and the costs of false negatives are greater than the costs of false positives over evolutionary history, selection will favour a bias towards false positives (Galperin and Haselton, 2012). Hence, and as its name suggests, the HADD device is calibrated in a hypersensitive way, leading to many false positives.

This hypothesised mechanism seems to enjoy a degree of empirical corroboration. Workman and Reader (2008) cite the rather striking example of a study by Fritz Heider and Mary-Ann Simmel in which participants attributed agency to two triangles and a circle depicted in a film as moving in and out of a box (the larger triangle, apparently, was described by one participant as an aggressor).

Other cognitive systems would also play their part. The systems responsible for us possessing a theory of mind would obviously also be crucial. By the age of three or four years, most children possess an ability to infer people's state of minds. The survival value of this is obvious, enabling us to predict the actions of animate objects in the world.

A wide range of unexpected events and ambiguous events trigger HADD and thereby are perceived to be invested with minded agency. As the causes of many of natural events that trigger HADD—such as violent earthquakes, thunderous storms and the like—are unseen, in a prescientific world it is a natural move to ascribe them to supernatural agents. These supernatural agents are counterintuitive, which makes them attention-grabbing and memorable. However, as Boyer (2002) stresses, these beliefs are *minimally* counterintuitive: beliefs that are easy to remember (in virtue of being counterintuitive) and easy to use (thanks to agreeing largely with people's expectations). Although they violate our natural expectations of the world, they are not too 'exotic', as Bulbulia (2004) puts it, or too divorced from the world to escape our interest and attention.

Thus, we arrive at the by-product of evolutionary understanding of religion: a hyperactive agency detection mechanism feeds into theory of mind and other mechanisms to enable us to 'see' a world imbued with supernatural agents that are minimally counterintuitive, which is then culturally mediated in terms of poly- or monotheism together with other supernatural agents (e.g. angels, which are practically interchangeable with the gods of polytheism).

The by-product approach is perhaps the most promising evolutionary psychology approach to religion. It is certainly a reasonable hypothesis that seeing supernatural agency operating in the world is partly due to this perceptual bias. The human experience of the world is as rich in false positives of supernatural agents as it is in false positives of natural agents. In a way, it would be extraordinary if the two were not connected. HADD therefore seems to have played a key role in the formation of beliefs about supernatural agency. Unlike the false positives of natural agency, the false positives of supernatural agency can be very costly when it results in regular ritualised behaviour, but not so costly as to outweigh the tremendous fitness increases brought about by the HADD adaptation.

If religious belief and practice is a by-product of a convergence of several cognitive systems, then this leads to some novel predictions. For example, the failure of one of these systems in an individual should reduce the likelihood that the individual in question will forge or maintain religious beliefs. So people with autism, who essentially lack a theory of mind, should be less religious than average. And, indeed, this has recently been supported. Caldwell-Harris *et al.* (2011) was the first systematic study of the religious beliefs of people with high-functioning autism and Asperger's disorder. They found that people 'with autistic spectrum disorder were much more likely than those in our neurotypical comparison group to identify as atheist or agnostic' (*ibid.*: 3362) This approach also lends itself to the prediction that the most widespread conceptualisations of supernatural agents will be those that are minimally counterintuitive, supernatural agents that are conceived as having intentions, desires and agendas, agents that are neither too exotic nor too abstract.

3.3 The Two Research Strategies in Tandem

So far, we've been representing the methodology of evolutionary psychology as comprising two distinct research strategies: reasoning from form to function, and reasoning from function to the present. This is how Cosmides and Tooby present the methodology (e.g. Barkow *et al.*, 1992), how other evolutionary psychologists represent the methodology (e.g. Pinker 1997b) and how commentators, sympathetic and sceptical, represent it (e.g. Richardson 2007; Laland and Brown 2011a). I contend that evolutionary psychology's methodology is more sophisticated than the impression given by this presentation. This standard presentation misses out, or rather masks, something crucial. While these research strategies can be distinguished and used independently of one other, they can be used together, a point frequently missed in the literature, especially in the sceptical literature (e.g. Buller, 2005; Richardson, 2007), which tends to focus on indeterminacy problems arising from reasoning exclusively from form to function, or reasoning exclusively from function to form, but fails to consider the possibility that these strategies can be used in tandem.

Indeed, I contend the following: adaptationist analysis in psychology should utilise both strategies when formulating hypotheses. Furthermore, I suspect they often do, though I do not need to establish this. One can start at either end of the methodological chain between function and form, between past and present, but the point is to move up and down. For example, suppose we begin our adaptationist analysis with an extant psychological mechanism. We reverse engineer it—we conjecture that it solves an adaptive problem. But we don't end the analysis there. Once we've conjectured a selectionist scenario, we make use of the predictive approach, of moving up the methodological claim, making novel predictions about the design features of the extant psychological mechanism. Machery (forthcoming in *Oxford Handbook of Philosophy of Psychology*) articulates this as evolutionary psychology having a 'bootstrap strategy', using knowledge of extant psychological traits to hypothesise past selective pressures, and then hypothesising from these selective pressures to produce novel predictions about extant or new psychological traits.

Evolutionary psychology's proposal of sex differences in sensitivities to different forms of infidelity is a simple but illustrative example of the kind of point I'm making. A useful definition of jealousy is an emotional state 'aroused by a perceived threat to a valued relationship or position and motivates behaviour aimed at countering the threat' (Daly *et al.*, 1982: 12).

As this psychological trait bears hallmarks of being an adaptation (cross-cultural, functional specificity, etc.), evolutionary psychologists can reverse engineer it. Reverse engineering might yield the following reasoning: (i) the possibility of partner infidelity was a recurrent adaptive problem; (ii) in the EEA, jealously would motivate behaviour that would have discouraged, limited or prevented infidelity; (iii) those with the trait of jealousy would have enjoyed a distinct fitness advantage over those who didn't. It would be adaptive. Hence, the trait could have been favoured by natural selection, and so preserved, refined and spread.

This leads to understanding jealousy as an adaptation functioning to solve the adaptive problem of counteracting threats to a valued relationship. More precisely, Buss conjectured that,

> Jealousy (1) is an emotion designed to alert an individual to threats to a valued relationship, (2) is activated by the presence of interested and more desirable intrasexual rivals, and (3) functions, in part, as a motivational mechanism with behavioral output designed to deter 'the dual specters of infidelity and abandonment' (Buss and Haselton 2005: 506).

So we've moved down the methodological chain, reasoning from form to function. But, crucially, we can move up the methodological chain again, reasoning anew from function back again to form. Hypothesising about selection pressures in the EEA led Buss to a further hypothesis: that there should be sex differences in jealousy. He predicted that:

> Men and women differ psychologically in the weighting given to sexual and emotional cues that trigger jealousy, such that (i) *men*

> *more than women* become upset at signals of sexual infidelity ... and (ii) *women more than men* become upset at signals of a partner's emotional infidelity (*ibid.*: 506, original emphasis).

Men and women faced different adaptive problems, and hence different selective pressures, with respect to infidelity. A man's fitness risks being severely compromised if his partner engages in cuckoldry. If the cuckoldry leads to offspring, and the man remains in the dark about the cuckoldry, the man would be spending his precious resources on another man's child. In contrast, in the ancestral past a woman's fitness risks being severely compromised if her partner diverts resources to another woman. Her man could have multiple sexual liaisons with women, but this fact alone wouldn't threaten her fitness; only if he became emotionally involved with another woman would her fitness be at stake. Hence, sexual infidelity is a greater threat to men than it is to women, while emotional infidelity is a greater threat to women than it is to men. Hence, sexual infidelity should upset men more than women and emotional infidelity should upset women more than men.

More precisely, evolutionary psychologists have postulated at least 13 sex-differentiated design features (*ibid.*). These include the following novel predictions. Relative to men, women demonstrate greater memory recall of indications of emotional infidelity. In contrast, relative to women, men demonstrate greater memory recall of indications of sexual infidelity. Furthermore, men find it harder to forgive a sexual infidelity than they do to forgive an emotional infidelity and are more likely to end a relationship following the discovery of infidelity if the infidelity is a sexual infidelity rather than an emotional infidelity.

Then the final step: the various novel predictions were tested. In a number of studies, predicted sex differences in jealousy have been experimentally supported (e.g. Buss *et al.*, 1992, 1999; Shackelford *et al.*, 2002; Becker *et al.*, 2004) and continue to be supported (e.g. Miller and Maner, 2008; Schützwohl, 2008).

So having begun reasoning from the present to the past, from form to function, we need not stop there, we need not only reverse engineer what the psychological trait could have been selected for—we can also reason from the conjectured past to the present, to predicting novel design features of the already known trait. This has two consequences. First, even accommodation of some given trait can lead to novel predictions. If psychological trait *T* functions to solve adaptive problem *X*, then we should find certain basic design features in trait *T* for solving problem *X*. Second, a principle emerges: for established psychological traits, traits we already know about, accommodation precedes prediction. We'll return to this principle later in the chapter, when it'll be brought to bear on a recent challenge to evolutionary psychology. Furthermore, in moving up and down the methodological chain, we might also generate novel predictions about the design features of other extant psychological mechanisms that are required for the mechanism under analysis to solve the adaptive problem—or even predict new mechanisms entirely.

Moving down and up, or up and down, the methodological chain, between ultimate and proximate questions, is not a one-shot event. It is not a once-and-for-all event. It should be seen as a continuing pattern of activity. One can move up and down the methodological chain multiple times, sometimes in small ways, sometimes in large ways. Moving up and down the methodological chain allows the instigation and integration of new research. A hypothesis can lead to a cascade of further research, some trying to corroborate it, others trying to undermine it, and the hypothesis can be repeatedly revised until the data harden—or it can be abandoned. For example, one can make a claim that some physical feature will be attractive across cultures. Cross-cultural research would then be trigged, perhaps offering corroboration from some cultures, perhaps also notable exceptions. The exceptions might well signal that the proposed adaptation has an important developmental calibration. For example, perhaps a society that fails to fit expectations has lower or higher levels of parental investment than the other cultures. Hence, it might well be that the calibration of the adaptation is sensitive to parental investment. Conjecturing why selection would have favoured such a design feature can then give us a better picture of the EEA of the adaptation, which, in turn, could lead to further novel predictions about other design features of the trait (and, possibly, other traits).

Notice the implication of this: when first formulated, evolutionary psychology hypotheses are not the last word on a given trait, but the first step in uncovering its design, features of the trait that non-adaptationist perspectives are blind to. Hence, crucially, although evolutionary psychology hypotheses might start off as 'simple', they can progressively become more complex, progressively mirroring the adaptations they're seeking to model.

Those with a naive falsificationist temperament might start choking at this, lambasting that such hypotheses can always be adjusted to make them fit ever increasing waves of inconsistent data. As such falsificationists are an endangered species, we had better reassure them. As we saw in Chapter 2, adaptations are exquisitely designed, capable of generating great variability in output. Hence, it is only to be expected that initial evolutionary psychology hypotheses, based on limited data samples, will probably be capable of further modification and refinement. This is not some shoddy move to save the hypothesis from falsification. Even refined hypotheses should generate further novel predictions—and if a hypothesis is just continually being reworked to save it from an avalanche of awkward data, continually failing in its novel predictions, then it'll be abandoned. This, I submit, is no different to any other hypotheses-driven empirical science.

3.4 Heuristics for Identifying Adaptive Problems

On what basis can we identify possible adaptive problems? On the basis of which sources do we make such hypotheses? Four sources of evidence or ways of thinking about possible adaptive problems present themselves: (1) historical information already obtained from a range of disciplines such as archaeology and evolutionary anthropology; (2) design information from existing knowledge of apparent design features of psychological mechanisms; (3) middle-level evolutionary theories; and (4) taxonomies of information-processing domains.

3.4.1 Historical Information

There are many historical details available that enable one to think strategically about possible and probable adaptive problems. A number of resources are available for reconstructing past ancestral conditions: the archaeological record, hunter–gatherer studies, evolutionary anthropology research and other evolutionary sciences. Indeed, these disciplines are dedicated to the inferential reconstruction of the past. Cosmides and Tooby note that those who work in those disciplines can use thousands of established facts to guide research. These include the fact that

> Our ancestors nursed, had two sexes, hunted, gathered, chose mates, used tools, had color vision, bled when wounded, were predated upon, were subject to viral infections, were incapacitated from injuries, had deleterious recessives and so were subject to inbreeding depression if they mated with siblings, fought with each other, lived in a biotic environment with felids, snakes, and plant toxins, etc. (Cosmides and Tooby, 1997b)

It's easy to be dismissive of these general features of our ancestral past but miss the point: many of the things we take for granted in post-industrial society, like finding food, finding a mate and so on, were serious adaptive problems; indeed, they remain adaptive problems, and they are known with a certitude to have been adaptive problems—problems likely, the evolutionary psychologist is betting, to require sophisticated solutions. It's a testament to the success of our adaptations that such a list might sound banal today. And evolutionary perspective makes us sensitive to adaptive problems we might otherwise be blind to.

We know that the size of the human body and the human brain increased from the early Pleistocene to around 300,000 years ago; this would have escalated energy requirements; some of the escalated energy requirements might have been offset by a decrease in gut size, which makes it more difficult to process low-energy, easily available food resources; these factors created pressure to develop more effective behaviours and tools for hunting and gathering high-quality food resources, which, in turn, increases pressure for enhanced social cooperation; escalated energy requirements also escalate the costs of reproduction, especially lactation, which further pressures the need for enhanced social cooperation (Fuentes, 2009).

And although we cannot be absolutely certain, it seems highly likely that step-parenthood was a recurrent feature of our ancestral past. The possibility of finding oneself with a child but no partner would have always been present. After all, a man could readily die from violence, hunting and disease, leaving his partner and offspring alone. A woman could die in childbirth, or as a result of violence or disease, again leaving her partner and offspring alone. And such possibilities represented opportunities: a woman whose mate has died might have many years of fertility remaining; a man whose mate has died might have status and resources valuable to a woman who has yet to be partnered. It seems almost inescapable that step-parental care obtained in the Pleistocene.

So we have a good idea of the kinds of recurrent conditions and pressures that our ancestors faced in the EEA. We also have a good idea of the kinds of conditions and pressures that they did not face: they did not encounter cities teeming with anonymous faces, formal political and legal institutions, and societies with massive and visible inequalities of material wealth; they did not encounter written language, mass communication and mass media; they did not encounter high technology, high medicine, contraception and paternity tests. They simply did not encounter many of the conditions and pressures that only arose with agriculture and industrialisation. Do not miss the significance of this: different social environments generate different interdependent decision problems. Had our ancestors over thousands of years experienced dramatically different social environments than the ones they actually encountered, their interdependent decision problems would also have been dramatically different, and, accordingly, we would expect a dramatically different arrangement of psychological mechanisms to have evolved.

So we have a good idea of the kinds of recurrent conditions and pressures that our ancestors did and did not face in the EEA. They might be general but they are sufficient to enable us to identify possible adaptive problems. And our knowledge of ancestral conditions is likely to only get richer and more detailed, thereby making the details of those recurrent conditions and pressures more visible. Indeed, as we shall see in the next chapter, outstanding inferences can be secured from quite limited data and importantly often these inferential reconstructions can be tested in various ways. Indeed, evolutionary psychology does precisely this: it makes inventive references, testable inferences from—but within its own restricted remit of psychology.

3.4.2 Design Information

Existing knowledge of design features of extant psychological mechanisms, as well as new design knowledge secured from successful evolutionary psychology hypothesis testing, can also provide an invaluable pathway into our evolutionary past. 'The functional design of an adaptation is the record of the salient, long-term environmental problem involved in the creation of the functional design. Thus, we can actually scientifically go back in time and discover the creative selection pressures that were effective in human evolutionary history' (Thornhill, 1997: 11).

Symons (2005) is an excellent example of existing psychological factors providing a powerful torchlight into conditions obtaining in our ancestral past. A striking feature of human courtship is the fear that an approach will be rejected. The thought of rejection hurts. The memory of rejection hurts. We take this for granted. But such a strong fear, which scuppers and sabotages many potential approaches, appears 'astonishingly dysfunctional' in today's world (*ibid.*: 257). Cities are populated by people we don't know, and, if we meet, unlikely to meet again by chance. In such a world, 'The potential benefits of propositioning an attractive member of the other sex, which include everything from a sexual fling to a lifetime mateship, would appear to vastly outweigh the potential costs, which seem to consist mainly of a small amount of wasted time' (*ibid.*: 257). Even if one doesn't pursue a (socially taboo or potentially even illegal) numbers game, approaching as many potential mates as possible, and instead limits approaches to those perceived to be receptive to being approached, the fear of rejection often remains.

Just as our strong and dysfunctional preference for fatty and sugary foods provides important clues about ancestral environments, so too our strong and dysfunctional fear of rejection might provide important clues about ancestral societal conditions. Symons points out that fear of mate rejection might possibly reflect mate rejection having heavier costs in the past than it does today. As hunter–gatherers, our ancestors lived in relatively small groups. In such a situation, the risk of one's sexual and courtship rejection becoming common knowledge might be significantly higher than it is today. If that were so, and if rejection in the past carried with it substantial costs such as a diminishing one's perceived mate value in the eyes of others, then mate rejection anxiety could be adaptive.

> On a modern university campus, with thousands of students and enormous scope for anonymity, Bob's anxiety at the prospect of hitting on Bobbi is, perhaps, 'irrational' in the sense that he has little to fear but fear itself; but the underlying motivational system may have been shaped by selection to function in an environment in which rejection had substantial costs (*ibid.*: 257).

Now, of course, if one were interpreting evolutionary psychology exclusively in terms of being an explanatory project, which is the typical interpretation of evolutionary psychology in the literature, then one might easily misconstrue such a move as being a vicious circle: that one projects an observed psychological trait back into the past and then simply projects it forward into the present and thereby consider the trait explained. However, when one looks at evolutionary psychology as a heuristic project, then one can see that projecting an observed psychological trait back into the past in this way enables us to further model past selection pressures and to consider possible further design features of that trait, which can then lead to psychological experiments being designed in a way to verify whether or not the trait has the hypothesised hitherto unknown design features (the bootstrapping notion mentioned earlier).

3.4.3 Middle-level Evolutionary Theories

Middle-level evolutionary theories, theories of specific selection pressures, theories such as parental investment theory, reciprocal altruism and sexual strategies theory, provide a further guide and inspiration for hypothesis generation. As Machery (forthcoming in *Oxford Handbook of Philosophy of Psychology*) notes, because they are broad, applying to the evolution of numerous taxa, middle-level theories say little about what specific traits might have been selected for by the selective pressures they specify.

Hence, middle-level theories alone do not offer straight deductions into hypotheses. But as Nancy Cartwright has pointed out over the years, even in physics, applications of a scientific theory are often not simple acts of deduction. Applying a theory to real-world situations takes acts of creativity; Nobel prizes are not awarded for just pressing a button on a conceptual vending machine, as if theory just shoots out applications (models) with a little deductive effort.

Although middle-level theories will not give us straight deductions, they can, nevertheless, help guide and build hypotheses. Recall an example from Chapter 1. On the strength of kin selection, which favours altruism between genetic relatives, Daly and Wilson (1980) proposed that we have a psychological adaptation for discriminative parental solicitude. Parents will tend to experience parental feeling for their natural offspring and will be moved to provide care for them. Unfortunately, stepchildren will be less likely to trigger this adaptation. Hence, the infamous 'Cinderella effect'.

The so-called 'good genes' view of sexual selection has helped generate a number of hypotheses and novel predictions. This view suggests selection has favoured mate preferences for healthy individuals owing to inclusive fitness benefits associated with mating with such individuals—if so, attractiveness judgements should reflect health judgements (Gangestad, 2000; Jones *et al.*, 2001).

Low fluctuating asymmetry is a possible visual marker of underlying health. Indeed, it might reflect an individual's heritable ability to maintain good health in the face of developmental shocks. During development, genetic perturbations, such as deleterious recessives and inbreeding, and environmental perturbations, such as reduced nutrition, disease infection, parasitic infection and toxicity exposure, can give rise to fluctuating asymmetries, deviations from symmetry in bilateral traits that are, on average, symmetrical at the population level (Swami and Salem, 2011). Individuals who are better able to maintain developmental stability in the face of such genetic and environmental perturbations tend to have low levels of fluctuating asymmetry. Low levels of fluctuating asymmetry are correlated with health (e.g. Waynforth (1998) found lower fluctuating asymmetry successfully predicted lower morbidity in a population sample in Belize), while higher levels of fluctuating asymmetry are associated with certain disabilities. Accordingly, selection might have favoured mate preferences for lower levels of fluctuating asymmetry.

This line of reasoning has led to a harvest of hypotheses and novel predictions on the relationships between symmetry and mating preferences. For example, high symmetry male faces are judged to be more attractive than low symmetry male faces (Perrett *et al.*, 1999). Men with high symmetry experience greater mating success than less symmetrical men, have sex at an earlier age and gain quicker sexual access to new partners (Gangestad and Thornhill, 1997).

Furthermore, as evolutionary psychologists have established, women's sexual preferences change across the ovulatory cycle. For example, during the fertile phase of their ovulation cycle, women particularly prefer more masculine male faces (Penton-Voak *et al.*, 1999; Penton-Voak and Perrett, 2000) and deeper male voices (Puts, 2005). On the strength of ovulatory cycle research and symmetry research, Gangestad *et al.* (2005) hypothesised that women possess an adaptation to be attracted to symmetrical men when near ovulation. From this adaptationist hypothesis, a novel prediction follows: the benefits of extra-pair mating with symmetrical men outweigh its costs only for women with high fluctuating asymmetry partners. Hence, the ovulatory cycle shift in women's extra-pair desires and flirtation should be strongest for women with high fluctuating asymmetry partners.

3.4.4 Taxonomies of Information-processing Domains

Selection pressures can be conceived as arising from social and ecological domains. Geary (2005) provides a valuable taxonomy of information domains: within the social information domain or folk psychology are the subdomains of self, individual and group, within which are further subdivisions; within the ecological information domain are the subdomains of folk biology and folk physics, again with various attendant subdivisions.

We can use taxonomies like this to identify possible adaptive problems: for example, within the social subdomain of group, we can think of adaptive problems pertaining to kin, in-groups, out-groups and group identity; within the ecological subdomain of folk physics, we can think of adaptive problems pertaining to movement, representation and tool use.

3.4.5 Further Refinements

Furthermore, we can refine adaptive problems in at least two ways: (1) possible sex specificity of the adaptive problem and (2) the grain of the problem. I shall say more about (2) later in the chapter. Regarding (1), men and women faced a number of identical adaptive problems: most obviously, they needed to source food and security, and detect cheaters. But they also faced different adaptive problems: some adaptive problems are unique to women and some unique to men. For example, women face adaptive problems relating to childbirth that led to particular physiological adaptations unique to women. Therefore, it's entirely possible that there are sex differences with respect to the configuration or even the existence of particular psychological mechanisms. Hence, we can ask not only whether there was a recurrent adaptive problem in some information-processing domain but also whether men and women identically faced this adaptive problem or not.

3.5 Heuristics for Identifying Adaptive Solutions

The engineering demand for function and form to be well-matched provides the central organising way of thinking about what kind of adaptive solution could have been selected for a given adaptive problem. To consider what a well-matched solution would be like is to consider the kind of cues to be processed as well as the kind of developmental and contextual calibrations that would constitute form–function match. The following specific heuristics guide evolutionary psychologists in their identification of adaptive solutions.

3.5.1 Cues

Psychological adaptations process information and therefore receive information or cues as inputs. There is a wide spectrum of cues: tactile cues, olfactory cues, audio cues, vocal cues, static visual cues, dynamic visual cues and so on. People unconsciously leak or broadcast cues all the time, in their voices, in their movements, in their eyes, in their odour, and such cues can have a surprising amount of reproductively relevant information. For example, dance reveals symmetry, especially in young men (Brown *et al.*, 2005). There are vocal cues of ovulation in human females (Bryant and Haselton, 2009). And during the fertile phase of their ovulation cycle, women particularly prefer the scent of men who are low in fluctuating asymmetry, which, as discussed previously, is a marker of heritable fitness (Gangestad and Thornhill, 1998; Thornhill and Gangestad, 1999).

Psychological adaptations might be designed to receive surprising information and cues—surprising in the sense that we'd ordinarily not consider the cue type to be relevant to the adaptive problem at hand. Therefore, taxonomies of cues can be heuristically powerful because they can enable us to think of different types of input than we are accustomed to thinking. For example, research on disgust usually focuses on visual and olfactory detection cues. Oum *et al.* (2011) investigated whether tactile cues might also play a role in pathogen detection. In their study, participants briefly touched and then rated stimuli. Results show that participants rated stimuli resembling biological consistencies as more disgusting than stimuli resembling inanimate consistencies, suggesting that tactile cues provide information for disgust-related processes.

Furthermore, taxonomies of cues enable us to not only think of different types of cue a psychological might receive, but can also enable us to explore a range of cues within an existing category of input. For example, visual cues can be subdivided into movements; shapes (specific shapes, specific curves, convex, concave); sizes; colours; lighting and shades; contrasts; brightness; sharpness; and location within the visual field (left, right, top, bottom, background, foreground, two dimensional, three dimensional). And auditory cues can be subdivided into types of sound (human, animal, ecological) and qualities of sound (volume, tone, tempo, rhythm, duration). So even if we have good reasons to believe that a purported psychological mechanism receives only one type of cue such as visual cues, it might process in elaborate ways a far greater spectrum of that type of cue than is currently known—or will be known, unless we think strategically in the manner recommended by these heuristics.

3.5.2 Developmental Calibrations

Evolutionary psychologists usually focus on functional questions about adaptations, and focus less on the ontogeny of adaptations (but without denying the importance of development). But does this mean developmental perspectives are irrelevant to evolutionary psychology hypothesis generation? Absolutely not. Thinking about developmental processes can greatly enrich evolutionary psychology hypotheses. Although adaptations are held to be developmentally robust, their calibration can be developmentally sensitive. In animals, early life experiences modulate behaviour. More precisely, animals often adjust their life histories in response to early environmental conditions, in ways that enhance reproductive fitness, at least in the environment of evolutionary adaptation. For example, Frederick (2012) cites Cameron *et al.* (2004), who reported that high stress early environments lead female rat offspring to mature faster, become more sexually receptive and reproduce more quickly than they would otherwise. Likewise with humans: differences in the early physical (including prenatal) and social environments can play a critical role in the calibration, in the sensitivities, of particular psychological adaptations.

This is much like weather forecasting: when we go out and need to know what the weather will be like later in the day, we will gather information from some sources to make a best guess, whether from the weather report or simply looking outside the window, and to thereby adapt our attire accordingly. And likewise, some of our psychological adaptations might well be configured to scan early environments for forecasting clues to as to what kind of environment is likely to obtain at a later stage of development. Such cues might include prenatal toxin and disease exposure, postnatal toxin and disease exposure, and the quality and quantity of interaction with family and societal peers. Belsky *et al.* (1991) investigated the impact of early environmental cues on the development of mating strategies. They hypothesised that an early environment characterised by relatively short and unstable relationships and scarce and unpredictable resources should predict differences in individual propensities to engage in short-term versus long-term mating strategies.

> Individuals whose experiences in and around their families of origin lead them to perceive others as untrustworthy, relationships as opportunistic and self-serving, and resources as scarce and/or unpredictable will develop behavior patterns that function to reduce the age of biological maturation ... accelerate sexual activity, and orient them toward short-term, as opposed to long-term, pair bonds ... Individuals, in contrast, whose experiences lead them to perceive others as trustworthy, relationships as enduring and mutually rewarding, and resources as more or less constantly available from the same key persons will behave in ways that inhibit (relative to the first type) age of maturation, will defer sexual activity, and will be motivated to establish—and be skilled in maintaining—enduring pair bonds, all of which will serve to enhance investment in child rearing (*ibid.*: 650).

These early environmental cues influence the development of mating strategies in ways that would be adaptive to the type of environment signalled by those cues: 'In essence, we argue that early experiences and the psychological and biological functioning they induce lead individuals to engage in either a "quantity" or a "quality" pattern of mating and rearing' (*ibid.*: 650).

In other words, Belsky *et al.* (1991) postulated the hypothesis that 'a principle evolutionary function of early experience' is to harvest understanding of the available environment, of the trustworthiness of others and so on. Again, this follows the logic of adaptationist hypothesising in psychology: if *X* functions to do *Y*, then it should have certain design features. In this case, if the function of early experience is to harvest environmental information, this should lead to alternative developmental calibrations. As long-term mating strategies will tend to be less effective in unstable versus stable environments, early cues that signal that the environment is unstable should lead to a propensity for short-term mating strategies and vice versa for the signalling of a stable environment.

Indeed, there's a growing literature investigating whether early cues signalling a stressful or suboptimal environment might lead an individual to developing a strategy favouring short-term gains and early reproduction. For example, Nettle *et al.* (2010) report that among British women being born small for gestational age increased the likelihood of early reproduction. And Frederick (2012) found that birth weight predicts scores on the Attention-Deficit/Hyperactivity Disorder Self-report Scale, as well as attitudes towards casual sex, in college men.

3.5.3 Contextual Calibrations

Good design often demands good flexibility. As we saw in Chapter 2, adaptations have decision rules that evaluate environmental inputs and produce different responses accordingly. Hence, evolutionary psychology can make context-sensitive predictions about the operation of psychological mechanisms.

For example, concerns about levels of attractiveness in potential mates should be more advantageous for those living in environments subject to high pathogen levels than for those living in environments subject to lower pathogen levels. On the strength of this, Gangestad and Buss (1992) investigated the relative importance of mate attractiveness in 29 cultures. As predicted, they found that people living in high-level pathogen environments valued attractiveness more than those living in lower level pathogen environments.

More recently, Buss and Duntley (2011) examined the context-dependence of intimate partner violence. Buss and Duntley hypothesise that intimate partner violence is often directed towards solving one or more of nine adaptive problems: mate poaching, sexual infidelity, mate pregnancy by an intrasexual rival, resource infidelity, resource scarcity, mate value discrepancies, stepchildren, relationship termination and mate reacquisition. Buss and Duntley suggest that violence will be selectively deployed in ways highly contingent on personal, relationship, social, economic and cultural conditions. 'The key point is that an evolutionary lens has heuristic value for predicting the circumstances in which intimate partner violence is likely, and even the particular forms it is likely to take' (*ibid.*: 415).

3.6 Testing Predictions

Novel predictions pertain to outcomes found in experimental settings and so it's worth briefly looking at the range of experiments evolutionary psychologists can deploy in order to confirm or disconfirm predictions. Evolutionary psychologists are, unsurprisingly, psychologists, and hence they use, devise and execute standard psychological experiments to confirm or disconfirm their novel predictions.

A standard method of confirming or disconfirming people's predicted responses and preferences is to survey them, whether by questioning them or giving them questionnaires. Such questionnaires can be crafted to investigate short-term mating preferences using various variables. For example, in order to test a prediction that men are more inclined to novelty than females, Symons and Ellis (1989: 133) crafted the following question:

> If you had the opportunity to copulate with an anonymous member of the opposite sex who was as physically attractive as your spouse but no more so and as competent a lover as your spouse but no more so, and there was no risk of discovery, disease or pregnancy and no chance of forming a durable liaison, and the copulation was a substitute for an act of marital intercourse, not an addition, would you do it? (Symons and Ellis, 1989: 133, as quoted in Cartwright, 2008: 250).

Questionnaires can also be crafted to investigate long-term mating preferences. For example, people can be asked to rank long-term mate characteristics, such as 'good financial prospects', 'intelligence' and height', identify minimum characteristics required of a long-term mate, or identify ideal characteristics in a long-term mate.

Creating and administrating questionnaires has obvious strengths: the tests are simple and relatively cheap, making them expedient to use cross-culturally. However, questionnaires have an obvious weakness. It's a commonplace that people's stated preferences don't necessarily match their actual preferences. Someone might state a preference for such and such characteristics, but, in fact, a different set of characteristics or even entirely opposite characteristics might trigger attraction. This mismatch between stated and actual preferences can be due either to lack of personal understanding, often due to personal identity narratives rationalizing away certain preferences, or owing to an expectation to conform to certain social norms or stereotypes. This is not to say that self-reports have no utility, but that they should be used with care. And, where possible, other types of test should be utilised to test predictions.

In attraction research, another way to test preferences is to investigate personal dating adverts in which people describe both their own characteristics and the characteristics they seek in a potential mate (Pawlowski and Dunbar, 1999a, 1999b; Pawlowski and Koziel, 2002; de Sousa Campos *et al.*, 2002). One advantage of investigating personal dating adverts for information on mating preferences is that such resources are unlikely to have an expectation bias skewing stated prefaces: these dating adverts are often anonymous. Furthermore, the adverts represent an investment of effort, and thus reflect actual attempts to find actual partners. Nevertheless, personal dating adverts still have drawbacks. The constituency of those who use dating adverts—those who submit and respond to such adverts—might be unrepresentative of the population at large. Furthermore, stated preferences in dating adverts can again diverge from actual preferences, not because of expectation bias, but owing to personal identity narratives rationalising away certain preferences

An alternative to investigating self-reported preferences is to investigate actual short- and long-term mating, relationship and activity patterns. This can be done by mining statistical data and archives. For example, if males prefer females who are younger than them and females prefer males who are older than them, this should be mirrored in marriage statistics. Fast-moving technological developments might elevate digital archival data as an increasingly importance source for testing evolutionary psychological hypotheses. For the first time in history, vast amounts of social data exist in searchable devices. Such endless seas of data are ripe for hypotheses making predictive waves over them.

Furthermore, rapid technological developments enable the customisation of experimental parameters in ways that were simply not possible even just a few years ago. Indeed, Miller (2012) advocates engaging with smartphones seriously as research tools. Miller believes that smartphones have the potential to not only collect vast amounts of data relevant to evolutionary and non-evolutionary psychology from large and diverse constituencies around the world, but to also make such data relatively easily available to researchers. 'If participants download the right "psych apps," smartphones can record where they are, what they are doing, and what they can see and hear and can run interactive surveys, tests, and experiments through touch screens and wireless connections to nearby screens, headsets, biosensors, and other peripherals' (*ibid.*: 221). Miller provides numerous examples where such technological developments can provide new opportunities for testing novel predictions. For example, Haselton and Gangestad (2006) predict husbands will increase mate guarding activity when their wives are most fertile. Call logs could give firm data on whether husbands are, indeed, performing increased mate guarding activity when their wives are most fertile—for example, by texting and calling their wives more than they would otherwise. Another example is that GPS data could reveal whether peak-fertility women go out more often to bars and clubs, in line with evolutionary psychology predicting that females at peak fertility increase mate searching (*ibid.*).

A further method of confirming or disconfirming people's predicted responses and preferences is to use physiological measures. When asked to imagine either sexual or emotional infidelity on the part of a mate, Buss *et al.* (1992) found that men showed elevated heart rate and electrodermal activity to imagined sexual infidelity than to imagined emotional infidelity. The reverse pattern obtained for women. Instruments exist to measure facial muscle activity, allowing the collection of a large body of reliable empirical data on facial expressions (Schmidt and Cohn, 2001). Furthermore, on the strength of prior research, which established a positive relationship between men's testosterone levels and infidelity (in particular, that men with relatively high testosterone levels report sustained interest in sex beyond their current committed relationship, a greater number of sex partners, and a higher number of extra-marital affairs), O'Connor *et al.* (2011) predicted that women would attribute high infidelity risk to masculinised men's voices. Testing this prediction was possible in virtue of the PSOLA method (Pitch Synchronous Overlap Add), a standard technique of voice manipulation that selectively manipulates frequencies and harmonics.

3.7 Methodological Challenges Revisited

Now that we have a richer understanding of evolutionary psychology methodology and how it is used to generate novel predictions, let's revisit four of the methodological challenges we originally saw in the previous chapter and see how they now stand: the no stable problems objection, the fine grain problem, the no constraints objection and the issue of whether evolutionary psychology really is heuristic in practice. Regarding the other methodological objection we looked at in the previous chapter, the thinness of evolutionary psychology explanation, we shall examine that in greater detail in the next chapter, where among other things we shall discuss the issue of whether it's possible to make adaptationist explanations deeper.

3.7.1 No Stable Problems Objection

Recall that Sterelny (1995) claims that there are, in fact, no stable problems to which natural selection can engineer specialised solutions in the manner proposed by evolutionary psychology. Recall that the thought is that evolutionary arm races destabilise adaptive problems and thereby destabilise evolutionary psychology.

> So there will be real troubles in store for a methodology of discovering the mechanisms of the mind that proceeds by first trying to discover the problems which it must solve, and then testing for the presence of the solutions. For that methodology does not reflect the interactive character of social evolution (*ibid.*: 372).

We can now see that Sterelny has made an incorrect inference. The consequence of evolutionary arm races is not that adaptations cannot evolve; rather, it means we should expect them to be richly calibrated, as well as to expect the coevolution of antagonistic adaptations. The more two evolutionary opponents, such as males and females, can second-guess each other, the more complex their relevant psychological adaptations become. Just as it seems inescapable that organisms are well adapted to particular environments, so too it seems inescapable that certain adaptations are reinforced and made more complex by repeated bouts of competition with antagonistic coevolving rival mechanisms.

Furthermore, Sterelny is wrong to claim that evolutionary psychology cannot reflect this interactive character of evolution. It can and does. As before, what Sterelny sees as a problem for the programme is actually a heuristic strength. For example, human females have concealed ovulation. Thanks to evolutionary psychology, we now suspect, and have some evidence, that men can unconsciously detect ovulation cycles. During fertile periods, women unconsciously increase mate search activities; and men unconsciously counter-act by increasing mate guarding activities. It's a dialectical dance, a Darwinian dance, one performed daily below the thresholds of consciousness.

Duntley and Shackelford (2012) propose that an antagonistic, coevolutionary arms race has produced adaptations to strategically exploit others and defences to avoid the costs of victimisation. They hypothesise that adaptations to damage status co-evolved with victim defences against status damage; adaptations for theft and cheating co-evolved with victim defences against theft and cheating; and adaptations for violence co-evolved with victim defences against violence.

Indeed, one can postulate feedback within dyadic antagonistic coevolution. For example, Duntley and Shackelford (2008) hypothesise that adaptations that produce criminal behaviour create selection pressure for the evolution of counter-adaptations in victims, which, in turn, create novel selection pressures for the evolution of counter–counter adaptations in criminals. Similarly, Trivers (2011) hypothesises that because selection has led to deception detection, there has been selection for self-deception so as to better limit and hide deception cues from others. In other words, because people have adaptations to detect deception, we self-deceive ourselves in order to better deceive others.

Furthermore, we can propose not just dyadic antagonistic coevolution but also triadic antagonistic coevolution. Duntley and Shackelford (2008) note that when three individuals have conflicting interests in the same adaptive problem domain, a refinement in one individual's adaptation can simultaneously create new selection pressures on the other two individuals. The counter-adaptations that evolve in each of the other two individuals as a result can then create further selection pressures.

3.7.2 The Fine Grain Problem

Evolutionary psychologists hypothesise domain- or problem-specific adaptations. So there will always be a question of granularity. Recall the fine grain objection: How do we individuate adaptive problems? How do we characterise them? How fine or coarse is the grain of a domain? How specific is the adaptive problem?

The first response is born out of considerations articulated in this chapter: namely, to stress that these possibilities can be tested. Is predator threat a single adaptive problem, or one of a series of adaptive problems, with different predator threats creating different adaptive problems? Well, we can put these different possibilities to the test. These different options should lead to different design features. For example, a psychological adaptation dealing with predator threats in general should have different design features to a psychological adaptation just dealing with a subset of predator threats.

Hence, instead of being seen as an embarrassment, the issue of individuation, which often arises when refining a successful hypothesis, can be seen, and I believe should be seen, as a heuristic strength. For example, a recurrent adaptive problem is the possibility of female partner infidelity. The fitness costs for failing to solve this problem are cataclysmic; solving this problem will deliver significant fitness gains. Upon closer design analysis it becomes apparent that the adaptive problem of female infidelity is composed of a number of subproblems including: (1) preventing female infidelity; (2) correcting female infidelity; and (3) anticipating female infidelity (Shackelford, 2003). Each of these possibilities opens up new lines of research. Now notice this isn't an outrageous unconstrained speculation. This isn't going into labyrinths of possibilities and frustrating the usefulness of the heuristics: it's a simple and effective breaking down the problem into several subproblems more suitable for hypothesising and designing experiments over.

So the fact that an adaptive problem can often be decomposed into several subproblems and further refined can only be a good thing if we're on a quest of exploration and discovery.

The second response is to highlight that this problem is ubiquitous to functional hypotheses, not unique to evolutionary functional hypotheses. As Machery (forthcoming in *Oxford Handbook of Philosophy of Psychology*) notes, psychologists, who are in the business of investigating psychological traits, often characterise these traits functionally, and hence their efforts are equally subject to the problem of individuation. So really there is nothing new here.

3.7.3 No Constraints Objection

Even with good heuristics it's possible to produce more than one hypothesis for a given target and hence it is only to be expected that it's sometimes possible to hypothesise alternative adaptive solutions for a given adaptive problem. In the case of an evolutionary science this, especially, shouldn't be seen as surprising: there is a degree of contingency present in the evolution of traits and alternative evolutionary trajectories are possible. This also shouldn't be seen as problematic. If alternative adaptationist hypotheses for a given target are possible this should be articulated to the point of generating rival, mutually exclusive predictions across one or more sets of observable measurements. Laying out rival predictions, specifying what would count as evidence for or against these rival hypotheses, finding the data and then evaluating the hypotheses—that's what a hypothesis-driven empirical science should do.

Evolutionary psychologists themselves are not blind to this. For example, Alcock and Crawford (2008: 37) claim that evolutionary psychologists recognise that sometimes several adaptive solutions can be hypothesised for a given trait and that they test them accordingly:

> These articles often have considered several different tentative hypotheses on the phenomenon, a reflection of the fact that adaptationist researchers can often think of multiple explanations for this or that trait. When there is more than one hypothesis to consider, the need for testing in order to reject incorrect ideas is obvious.

This is especially likely to be the case when an adaptive problem is being investigated for the first time. Buss and Shackelford (1997) is an illustration of this. They examined a range of mate retention tactics in marriage from an evolutionary perspective. They begin their paper by noting there has been little research on the adaptive problem of mate retention. One of several novel research questions they consider is mate retention behaviour in marriages where there is uneven mate value. One subquestion is whether men who marry women perceived to have higher mate value will dedicate more effort to mate retention than men married to women with equal or lower mate value than themselves. Adaptationist considerations strongly suggest they would. The other subquestion is whether women who marry men with higher mate value will tend to heighten or to relax retention efforts. Here adaptationist considerations can generate opposite hypotheses, as Buss and Shackelford freely and happily acknowledge.

One evolutionary psychology hypothesis that can be formulated is that women in marriages where the man is perceived to have the higher mate value will dedicate more effort to mate retention than women in marriages where the man is perceived to be as equal or have less mate value than the woman. This hypothesis focuses on the dramatic costs of losing the higher mate value partner to a competitor entirely and the gains to be made by successfully retaining the partner exclusively. The alternative evolutionary psychology hypothesis that can be formulated is that women in marriages where the man is perceived to have the higher mate value will dedicate less effort to mate retention than women in marriages where the man is perceived to be as equal or have less mate value than the woman. Here, an additional set of considerations is being factored in: higher value men tend to be capable of fathering children in different relationships simultaneously; higher value men might feel an entitlement to additional relationships; if a woman engages in significant mate retention behaviour with a highly prized male who desires to be in more than one relationship then there is a risk of losing that highly prized man entirely; it might be better to secure the partial attention and resources of a highly valued male than secure the full attention and resources of a less valued man.

Again, if one were approaching this from an explanatory angle, especially with a naive falsification temperament, one would be deeply suspicious of all this: it would look like evolutionary psychology is trying to explain both *X* and *not X*. It looks like evolutionary psychology is simply accommodating phenomena no matter what it turns out to be. However, when one correctly approaches this from the heuristics angle, what is happening becomes clear: this together with a battery of other research questions is a first shot at understanding what psychological solutions might have evolved to solve the adaptive problem of mate retention. The first shot will give us some leg up, something to work with, something we can work with further. There should be nothing scandalous or even remotely suspicious about that.

Nevertheless a reasonable concern might now arise. Evolutionary psychology is fertile. The trouble, one might venture, is that it's too fertile. Perhaps its strategies and methods can generate a large number and range of hypotheses for any given adaptive problem. There needs to be a reasonable limit on the number of hypothesised adaptive solutions for a given adaptive problem. If this is not so, if the permutations are too great, then focusing on adaptive problems and solutions will not successfully reduce and constrain research space. Evolutionary psychology's heuristic value will be wholly compromised.

Recall from the previous chapter that in the sceptical literature, this concern is often cashed out into a very strong position: evolutionary psychology hypotheses are judged to be unacceptably unconstrained. The concern, as we saw, is entirely understandable but the position that sceptics reach is untenable. In practice, evolutionary psychology's heuristics allow, at most, only a limited range of adaptive solutions to be generated for a given adaptive problem—certainly a range suitable for experimentation. Think of the adaptive problem of infidelity. According to evolutionary psychology's heuristics, a well-designed solution to this problem is jealousy. Perhaps a clever individual could use the heuristics to generate one or two alternative proposals, but certainly not a runaway list of possibilities. Indeed, if the heuristics allowed for an unconstrained multitude of proposals to be generated for a given adaptive problem, we would see this in the evolutionary psychology research literature. Likewise, regarding the finer details of how any proposed psychological adaptation is calibrated, several design possibilities might present themselves, but again the range is well within the acceptable parameters of experimentation.

Indeed, crucially, as function and form should be well-matched, adopting an engineering perspective allows one to identify quickly and discount a large number of possibilities. For example, Symons (2008) asks us to consider a situation where someone proposes that the human female orgasm is an adaptation designed to promote conception by enhancing sperm retention, and that the human female organism is designed to achieve this goal only when the organism occurs close to the time of the male partner's ejaculation. Adopting an engineering perspective on this purported adaptation, a psychological adaptationist might ask

> why a series of coordinated muscle contractions in a female body that mimic the muscle contractions of ejaculation in a male body would miraculously turn out to be well designed to achieve an entirely different goal. That is, shouldn't we expect the design of a female device whose function is to promote conception by retaining sperm to differ in important ways from the design of a male device whose function is to propel semen from the body, for the same reasons that we expect a device designed to pump blood to differ from one designed to digest food? (*ibid.:* 9).

Furthermore, adds Symons, selection would favour males who observe the following behavioural rule: 'thrust until your partner orgasms, then immediately ejaculate'. As the form and the function are not well matched, and, indeed, the prediction doesn't obtain, the heuristics rule out this proposal as a serious contender.

Sceptics freely talk about 'unconstrained speculation' but fail to cite any specific areas where evolutionary psychology hypothesis generation is running amok, nor do they make the rather obvious and deeply subversive move of using its heuristics to generate a runaway list of hypothesised adaptive solutions for some given adaptive problem. Imagine if they did that. Imagine if a paper was published that used the heuristics to generate hundreds of hypothesised adaptive solutions for some given adaptive problem, all of which were consistent with adaptationist concepts and methodology. Wouldn't that be a decisive blow against evolutionary psychology as a heuristic? It surely would be. So why hasn't it been done? That such a move is not forthcoming anywhere in the literature is, I believe, most telling.

Although there will be some bets that fail to pay off, and some dead ends, this is unavoidable. Heuristics reduce the size of the research space, but researchers nevertheless still need to rely on a measure of trial and error within the constrained space. The initial practice of evolutionary psychology spreads far and wide, accommodating as much psychological phenomena as possible. Some of the initial outreaches will be fruitless, others more promising. The initially promising outreaches provide us with some fixed points that can anchor adaptationist hypotheses. Further research questions can be raised and answered, testable questions about calibration, contextual sensitivities and the level of the grain, leading to an unrelenting examination of the design details that constitute psychological adaptations.

3.7.4 Predictions or Accommodations?

Recall that, unlike most critics of evolutionary psychology, Schulz (2011) is alert to the heuristic potential of evolutionary psychology. Nevertheless, although he acknowledges it's possible for evolutionary psychology to be heuristic, and he cites Csibra and Gergely (2009) as an example of one such instance, he contends that this is, in fact, very rare, and so we shouldn't overemphasise its importance:

> this point must not be overemphasised—in fact, far from being a common occurrence, heuristic applications of evolutionary theory in psychology are actually quite a rarity. While such occurrences do exist, as yet, they are still in a minority: *most* cases of evolutionary psychological research—and, in fact, virtually all of the work of the Santa Barbara ('EP') School of evolutionary psychologists – employ evolutionary theory only to *explain* a known set of phenomena, not to lead us to *discover* these phenomena' (*ibid*.: 232; original emphasis).

This is a big empirical claim to make. Yet in his paper, Schulz considers only a few examples. So how can such a case be made? Schulz's strategy is straightforward: first, debunk the flagship example of past-to-present novel prediction as actually being a case of present-to-past accommodation; second, point out that a lot of well-known evolutionary psychology hypotheses are also accommodations of social phenomena we already know; third generalise this into a general claim about evolutionary psychology practice. So while evolutionary psychology in principle could be heuristic in practice it mostly churns out accommodations. Plausible, intriguing adaptationist rationales for a wide spectrum of social phenomena—but still accommodations. Or does it? In fact, this generalisation needs to be challenged, and it is time to do so.

The first response to make is perhaps the most obvious: the cases Schulz identifies as accommodation—Cosmides and Tooby's, Buss's and Pinker's work from the early 1990s, and Symons's, and Daly and Wilson's work from the 1980s—date from evolutionary psychology's early days. As the history and philosophy of science has taught us to expect, early output of a research programme is usually primarily accommodation. Evolutionary psychology follows the usual arc, the usual career curve, of scientific endeavour. Exploring what facts a research programme could conceivably cover is what we might expect a new research programme to do; it seems perfectly reasonable to see how far adaptationist theorising can stretch.

Evolutionary psychologists are not—and should not be—afraid to spread their canvas wide, to explore where adaptationist analysis can succeed. For example, some evolutionary psychologists have hypothesised that rape is a male adaptation (Thornhill and Thornhill, 1992; Thornhill and Palmer, 2000). Leaving aside the sensitive nature of the topic, and the controversy surrounding purported policy implications (something that has clearly antagonised sceptics, as we'll see in Chapter 5), one could argue that existing evidence for this hypothesis is weak and that purported adaptive benefits of rape can be won by other traits and strategies (Dupré, 2001; Kitcher and Vickers, 2003). True enough, but this misses something crucial—namely, the hypothesis is exploratory. Hagen (2004) points out that there is currently insufficient evidence to decide whether rape is a male adaptation or not—at this stage, the more appropriate question is whether rape can be plausibly conceived as a male adaptation. Hagen believes the answer is clearly yes, and that this should motivate efforts to conduct further research, to seek out and establish further lines of evidence.

We shouldn't lose sight of the value of accommodation. When evolutionary psychology 'only' accommodates a phenomenon this alone often allows us to see the phenomenon in entirely new light. Evolutionary psychology accommodations can be impressive as they can make the ordinary extraordinary. Things we take for granted, as being obvious, as being given, actually have a history, a natural history. Many find these explanations worthwhile, especially given our justified commitment to evolutionary theory, even if such research has yet to be articulated to the point of testability.

Furthermore, accommodation can be radical. Many of us are too young to know a time when the very notion of evolutionary accommodations of behavioural and psychological traits surprised and shocked people. Many of us weren't around when the early sociobiologist pioneers were subject to assault and battery (Segerstråle, 2001). These days, people have become acclimatised and receptive to the notion of evolutionary explanations of behaviour (though, as the rape adaptation hypothesis shows, there is still plenty of scope for shock and strong emotions). What was once radical is now a common way of talking about the origins of behaviour. Indeed, purported evolutionary explanations of behaviour are frequently alluded to in various branches of academia, in television documentaries, in popular books, and even in pub chats (evolutionary speculation in the 'cocktail party mode' as Gould memorably put it). Nevertheless, there can be considerable social pressure not to notice the obvious. To some, it's obvious that stepchildren incur elevated rates of abuse and homicide; to others, perhaps ideologically conditioned in a certain way, this is not at all obvious. Hence, perhaps, why the so-called Cinderella effect is dismissed by Schulz (2011) as being obvious but dismissed by Buller (2005) as being flatly mistaken, as not a genuine phenomenon.

We can make a further point. Recall a point made earlier in the chapter: often, in order to generate new predictions about some extant phenomenon we need to first accommodate that phenomenon. When learning a new language, we tend to begin by translating what we already know into the new language. My name is Andrew. *Mon nom est* Andrew. At this stage we're more concerned with getting things right rather than being creative. Only later when we are more advanced can we use the new language to explore directly new ideas not originally thought of in the original language. And, of course, to know a new language is to enable one to explore a new terrain in ways otherwise not possible, to open up possibilities closed to those unable to speak the language.

A research programme's heuristic power is revealed over the medium- and long-term, not immediately in its infancy. It is only revealed in its entry into maturity. Hence, we should foster a degree of tolerance towards young programmes, those chiefly accommodating phenomena into their explanatory framework. We must allow evolutionary psychology to first conceive, then nurture and develop hypotheses. Establishing what phenomena can plausibly and promisingly be accommodated within evolutionary psychology's conceptual and methodological framework can take some time. However, as accommodation efforts wind down, prediction efforts energise, as per the heuristic framework I outlined above. Even if Schulz is right, even if evolutionary psychology is currently primarily the practice of accommodation, I venture that increasing waves of evolutionary psychology practice will focus more and more on generating novel predictions and devising elaborate tests for them. Liddle and Shackelford (2009: 291) put the point nicely: 'Theoretical articles are immensely useful in providing a framework for future empirical research on a particular topic. To dismiss an entire field because some hypotheses have not yet been tested is premature at best and disingenuous at worst'.

So there is no 'instant rationality', to borrow a memorable phrase from Lakatos, in the sense of an immediate judgement of a research effort. To properly appraise a research effort, especially one built for discovery, we must adopt a longer-term view. After all, a research programme that has hitherto been primarily in the business of accommodation might soon become stunningly predictive in stunningly novel ways.

We can acknowledge that some well-known evolutionary psychology hypotheses, such as the cheater detection hypothesis, are, indeed, accommodations of extant phenomena. However, as I argued earlier in the chapter, once accommodated, the framework of evolution psychology allows for novel predictions to be generated about the accommodated phenomenon. If this hasn't happened in some cases yet that doesn't mean it can't happen in those cases.

But a weightier response to Schulz's characterisation of evolutionary psychology is possible and it's time to develop it: despite cases of accommodation, there are clear examples of evolutionary psychology practice that delivers observational findings beyond what we already know, beyond the reach of the prevailing orthodoxy. And this is far more common than is generally known.

Earlier in this chapter I identified some of these: disgust as pathogen avoidance; cue detection of the ovulatory cycle; sex differences in jealousy; and female preference for symmetrical males. Indeed, perhaps inadvertently Schulz supplies the latter work as a further example of heuristic evolutionary psychology—though one has to sail through the footnotes to find this. After citing Csibra and Gergely (2009) as being, in his judgement, one of the few instances of evolutionary psychology being heuristic, Schulz appends a footnote, saying:

> Andrews et al. (2002, p. 538) and Buss et al. (1998, p. 545) claim that Thornhill & Gangstead's work on female preferences for symmetric men (see e.g. Gangstead & Thornhill, 1997) provides another example of a heuristic form of evolutionary psychology. Whether they are right in this is not something I shall discuss here (for some

> critical remarks concerning this, see e.g. Fuentes, 2002); what matters for present purposes is just that most instances of evolutionary psychological research are not heuristic in structure, and that finding exceptions to this requires hard work (2011: 227, n. 11).

Two observations. First, here is another example of heuristic evolutionary psychology practice. And yet it's exiled to the footnotes. Why is this? Surely this is more than a footnote? I think that it is. I think it should merit some attention. Second, we're told to not be too concerned about this case—what matters is that 'most instances of evolutionary psychological research are not heuristic' and it's been 'hard work' finding exceptions. But doesn't that sound a little like begging the question?

Sceptics never cease to question evolutionary psychology's output, suspecting it as being plausible but evidentially lacking accommodations, whereas I suspect its real history, especially its recent record, is full of surprising novel predictions gradually leading to bounties of new data.

G. K. Chesterton once remarked that if a job is worth doing, it's worth doing badly. When adaptationist hypotheses were first pursued in psychology, perhaps some of them were creaky, perhaps some of them were of questionable value. Again, we need not name any names. But evolutionary psychology is no longer in its infancy—though it's not quite entering middle age yet either. Reading evolutionary psychology research published in a variety of journals, as opposed to just reading second- and third-hand accounts of evolutionary psychology, what soon becomes apparent is that many papers table novel predictions and seek evidence for them, usually by constructing and conducting their own experiments. Of course, not every evolutionary psychology research paper does this, but those papers that don't hypothesise novel predictions still carry out important research, such as testing existing evolutionary psychology hypotheses in different contexts; testing existing evolutionary psychology hypotheses across cultures (e.g. Sznycer *et al.* (2012) examined cross-cultural similarities and differences in proneness to shame); finding fresh evidence for existing evolutionary psychology hypotheses (e.g. Ketelaar *et al.* (2012) provided new evidence for the existing evolutionary psychology hypothesis that smiles are associated with lower social status); and testing for, and ruling out, alternative explanations for successful novel predictions—exactly what one would expect of a hypothesis-driven empirical science. Indeed, it has been some time since I've read an evolutionary psychology paper that doesn't articulate its hypotheses to the point of testability or doesn't test existing hypotheses in different contexts or cultures.

And I think rather tellingly, the cases Schulz identifies as accommodation date from evolutionary psychology's early days, while the case he cites as demonstrating the heuristic potential of the research programme—Csibra and Gergely—is more recent, dating from 2009 (and the case he demotes to a footnote, Gangstead and Thornhill, is from 1997). Schulz cited the 1992 handbook *The Adapted Mind: Evolutionary Psychology and the Generation of Culture*, but for a more up-to-date presentation of the programme, it would have been better to consult and cite the 2005 *The Handbook of Evolutionary Psychology*. In that handbook, Buss notes, 'A decade ago, a handbook of this scope would have been impossible. The empirical corpus of research testing evolutionary psychological hypotheses was too slim. Now the body of work has mushroomed' (2005: xxiii). He further adds that, 'Hundreds of psychological and behavioral phenomena have been documented empirically, findings that would never have been discovered without the guiding framework of evolutionary psychology. Evolutionary psychology has proved its worth many times over in its theoretical and empirical harvest' (*ibid.*: xxiii). Yet Schulz focuses on a tiny sample of early examples of evolutionary psychology work, work done in the programme's infancy, when accommodation is only to be expected.

We can indeed point to a rich harvest of successful empirical predictions. Penton-Voak *et al.* (1999) and Penton-Voak and Perrett (2000) successfully predicted that women's preferences for masculine faces should be more pronounced during the fertile phase of the ovulation cycle than during non-fertile phases of the cycle. Haselton *et al.* (2007) successfully predicted changes in women's style of dress associated with ovulation. Gangestad *et al.* (2007) successfully predicted that higher fertility increases women's attraction to behavioural dominance. Lieberman *et al.* (2011) predicted that women near peak fertility in their cycle should reduce contact with male kin to minimise incest risks. Analysing itemised mobile phone bills from a number of women, they found that women at peak fertility talked less with their fathers, both in terms of number of calls and duration of class, but more with their mothers.

And that's just the tip of the iceberg of evolutionary psychology work on attraction and ovulation. We could cite many more examples of novel predictions from other areas of research. For example, Sell *et al.* (2009) advance a new hypothesis concerning the function and design of anger, which they call 'the recalibrational theory of anger'. According to Sell *et al.*, anger functions to orchestrate two bargaining tools—inflicting costs and withholding benefits—in interpersonal conflicts of interest, inducing the recipient of the anger to give greater weight to the welfare of the angry individual than they would otherwise, thereby helping resolve the conflict of interest in favour of the angry individual. Sell *et al.* postulate the existence of 'welfare tradeoff ratios' (WTRs):

> In social species, actions undertaken by one individual commonly have impacts on the welfare of others (measured in fitness, or in other currencies). Consequently, neurocognitive programs in social species should have been designed by selection to solve the following computational adaptive problem: For a given choice set involving self and other, how much weight should be placed on the welfare of the other compared with the self? We shall refer to the ratio of these weights as a welfare tradeoff ratio (WRT) between the self *(i)* and individual *(j)*: WTRij (2009: 15073).

Hence, as we interact with others, a neurocognitive programme is automatically calculating WTRs. If so, individuals could, in conflicts of interest, unconsciously use anger to influence others into recalculating their WTR. Individuals 'with enhanced abilities to inflict costs' (i.e. physically stronger individuals) or those with 'enhanced abilities to confer benefits' (i.e. attractive individuals) enjoy better bargaining positions when it comes to conflict of interest situations. Because of this, they will have 'a greater sense of entitlement' and hence are more liable to use anger in this way. Accordingly, they are more likely to prevail in conflicts of interest, seeing conflicts resolved in their favour. In total, Sell *et al.* tested 11 novel predictions they derived from their recalibrational theory of anger. They report that all 11 predictions were empirically corroborated and note that no other existing theory predicts the pattern of results obtained.

Interestingly, evolutionary psychology has also made successful novel predictions beyond social behaviour. DeBruine (2009) cites research illustrating how evolutionary psychology can generate novel predictions about non-social behaviours. It's been known for some time that humans estimate horizontal distances far more accurately than vertical distances. For example, when standing on top of building looking down, the distance people tend to perceive from the top of the building to the ground is overestimated. The distance estimated when standing on top of a five-storey building looking down is equivalent to the actual height of a nine-storey building. However, when orientated horizontally, people perceive the same distance much more accurately.

DeBruine points out that Jackson and Cormack (2007) make a further prediction: vertical distances should be overestimated more from the top than from the bottom. Jackson and Cormack propose what they call 'evolved navigation theory', which suggests perceptual and navigational mechanisms reflect navigational costs over evolution. Unlike ascent, descent involves a significantly increased risk of injury and death. In fitness terms, it's better to overestimate descent. Hence, we should find a descent illusion. The existence of this previously unknown descent illusion was recently confirmed in Jackson and Cormack (2007) and Jackson and Cormack (2008). DeBruine notes that this previously unknown descent illusion 'was predicted only by consideration of the fitness consequences of behaviour. Indeed, no one even thought to test this prediction for over 40 years until an evolutionary perspective was brought to bear on the topic' (2009: 932).

Consider the single case Schulz recognises as an actual instance of heuristic evolutionary psychology. Csibra and Gergely (2009) hypothesised that the capacity for natural pedagogy is an adaptation for acquiring generalisable local knowledge. If that is the function of natural pedagogy, then infants should be able to distinguish teaching episodes, where objective information is acquired, from teachers' personal tastes. The novel prediction was confirmed. Schulz is right to highlight this as a case of a heuristic evolutionary psychology in action—but, I believe, mistaken to think this is a special case. The logic underpinning Csibra and Gergely (2009)—if *X* is an adaptation for *Y*, then *Z* will follow—is exactly the logic at work in hundreds of other research papers, the logic outlined and discussed earlier in the chapter.

Schulz (2011) sought to examine whether evolutionary psychology is heuristic in practice but didn't articulate the heuristics of the research programme in any detail. Had the heuristics been articulated in detail I expect many more instances of its operation in practice would have become visible—and instead of seeing Csibra and Gergely (2009) as an outlier, a rarity, that case would instead have been seen as just another example of what one would expect from the heuristics in operational practice. In a way, it's not surprising the heuristics weren't explicated in detail: the heuristics are simply not well known or clearly or adequately captured in the literature. This theme of overlooking evolutionary psychology's heuristics—whether entirely or with respect to the logic and details of those heuristics—is an important theme, one that I return to in the final chapter.

So evolutionary psychology is heuristic in practice—to an extent far greater than commonly supposed. The case I set out to establish has now, I believe, been established: evolutionary psychology demonstrably has heuristic value—both in principle and practice.

3.8 Conclusion

Heuristics are expected to be both relatively constraining and to enable a moderate degree of creativity within the constrained space with respect to finer details. Furthermore, they are expected to yield a harvest of new knowledge. And this, I submit, is precisely what we see when we examine evolutionary psychology's heuristics with sufficient care.

We've identified examples of adaptationist thinking driving psychological research, research that delivers observational findings beyond what we already know, beyond the reach of the prevailing orthodoxy. Reasoning between ultimate and proximate questions, between function and form, asking what adaptive problems ancestral populations faced, considering what would constitute good design solutions to those problems, enables researchers to (1) discover new design features of extant psychological traits, and (2) discover hitherto unknown psychological traits. This is heart of evolutionary psychology. This is what adaptationist hypothesising can do for psychology.

The engineering demand for function and form to be well matched, the two broad heuristic strategies working in tandem and the specific heuristics—historical knowledge, design knowledge, middle-level evolutionary theories, information-processing taxonomies, possible cue inputs, and possible developmental and contextual calibrations—have enabled an impressive array of hypotheses and predictions to be formulated. They enable us to see possible design features that we would otherwise be blind to. They enable us to think structurally and strategically about design: In what domain does an adaptive problem obtain? Would men and women have faced this problem differently? How fine is the problem—does it decompose to further problems? What kind of contextual signals modulate the type and degree of behavioural output? And many more questions and subquestions. Initial evolutionary psychology hypotheses aim, or should aim, not for the last evolutionary word on a given phenomenon, but the first. They are in constant adjustment—both with the research programme's own findings and findings from adjacent research programmes and disciplines, a theme I develop further in the next chapter. If this is done, this should generate sophisticated hypotheses, as well as generate progressive increments to our understanding of psychological and social phenomena. And this is, of course, the normal course of science.

As evolutionary psychology operates primarily in the context of discovery, questions about indeterminacy of proposed adaptationist hypotheses are also rendered relatively unproblematic, as these alternatives can be tested too. If raising the issue of indeterminacy is to mean anything, it is to vigorously test a range of adaptationist hypotheses, not to abandon the research programme wholesale. The possible permutations of adaptive problems, solutions, and calibrations are not too numerous to investigate; they are well within the range suitable for testing and experimentation.

How good is the output of the programme? Is it primarily accommodation, or prediction? This is a question posed by Schulz, who, unlike many, recognises theoretical evolutionary psychology's heuristic value, but moves to deny that the discovery of new facts accurately characterises the practice of the research programme. In other words, to acknowledge that while theoretically evolutionary psychology has heuristic value, its actual practice falls far short of this. However, only a tiny sample of cases was considered—a tiny sample of old cases. Is this a sufficient basis on which to dismiss claims about evolutionary psychology being heuristic in practice? I do not believe so. Indeed, there is plenty of evidence that evolutionary psychology practice, especially more recent practice, is significantly predictive. I believe we can now say with reasonable confidence that evolutionary psychology as a research programme aims to discover new facts, that its theoretical and methodological framework allows novel predictions to be generated and tested, that this appears to be its rationale, its core, and that this happens in practice, significantly more frequently than is generally recognised. Thus, I contend, the case I set out to establish at the beginning of the chapter has now been established: evolutionary psychology demonstrably has heuristic value—both in principle and in practice.

4
Reframing Evolutionary Psychology as a Heuristic Programme

4.0 Introduction

We now have a richer understanding of what evolutionary psychology actually is in practice. The key is this: evolutionary psychology is a hypothesis-driven empirical science. The daily practice of evolutionary psychology is to focus on adaptive problems, hypothesise dedicated adaptive solutions and then subject these hypotheses to testing. This can generate novel predictions and there have been many instances where such novel predictions have been successfully corroborated, providing important new facts about features of our psychology. Hence the heartbeat of evolutionary psychology is that of discovery.

What would happen if we were to streamline the conceptual framework of evolutionary psychology? That is, what if we were to conceive evolutionary psychology as a heuristic project, as a simple but effective set of heuristic tools as outlined in the previous chapter, rather than as some kind of a paradigm with grand claims about the mind and the world?

In this chapter, we will explore streamlining the framework of evolutionary psychology—and the significant gains to be made by doing this. In 4.1, I contend that many of the tenets commonly associated with, or purported to be part of, the practical research programme are not strictly part of it. Accordingly, I will reframe evolutionary psychology's tenets, argue that this streamlining better matches evolutionary psychology's daily practice, and argue that this streamlining offers multiple benefits. As a result, much criticism about its purported tenets, such as massive modularity, simply ceases to apply.

In 4.2, I argue that this reframing clearly deflects many of the evidence criticisms of evolutionary psychology work—criticism that is the source of much unease and mistrust about the programme. For example, failure to establish all evidential particulars of an evolutionary psychology explanation does not in any way speak against evolutionary psychology when understood as a heuristic project.

In 4.3, I argue that framing evolutionary psychology as a heuristic project, rather than as a paradigm, enables us to better locate its unique place and role in the evolutionary and behavioural sciences. Establishing the multiple lines of evidence needed to vindicate fully evolutionary psychology explanations requires multidisciplinary activity. If criticism about the incompleteness of evolutionary psychology explanations is to mean anything, it is to stress evolutionary psychology's need for multiple disciplines across the behavioural and evolutionary sciences to engage with it. So while it's reasonable for sceptics to demand greater evidence for evolutionary psychology explanations, they're wrong in demanding evolutionary psychology alone satisfy such demands—these demands are properly allocated to the evolutionary and behavioural sciences collectively. The chapter ends with a moderate optimism concerning the prospect of the evolutionary and behavioural sciences collectively rising to the challenge.

4.1 Streamlining Evolutionary Psychology

In Chapter 1 we saw that the leading proponents of evolutionary psychology—Leda Cosmides, John Tooby, David Buss and others—present evolutionary psychology as a package of views:

- (T1) Evolutionary theory, with an emphasis on natural selection and adaptation
- (T2) The possibility of psychological adaptations
- (T3) Empirical adaptationism
- (T4) Inference from empirical adaptationism to massive modularity
- (T5) Methodological adaptationism
- (T6) Metatheory for psychology and the behavioural sciences
- (T7) Public policy agenda

To recap briefly: (T1) the modern evolutionary synthesis refers to the widely accepted bundle of theories that provide the standard account of evolution; evolutionary psychologists focus on natural selection and adaptations; (T2) is the idea that adaptations can be not only physiological but also psychological; (T3) empirical adaptationism holds that selection is the most important cause of most traits in most populations;

(T4) is the attempt to infer a strong form of massive modularity (that there are no domain-general psychological mechanisms) from empirical adaptationism; (T5) methodological adaptationism (a recommendation to approach psychological traits as adaptations—the strong form of this states that this is the best way to approach psychological traits, the moderate form states this is a useful or good way to approach psychological traits); (T6) is the championing of evolutionary psychology as a reformer and unifier of psychology and the social sciences; and (T7) is the championing of the application of evolutionary psychology research to public policy.

Evolutionary psychology as a research programme should not be identified with theoretical tenets surplus to requirement. It should not be identified with theoretical tenets that have no bearing on its daily practice. The key question here: Does a commitment to the search for psychological adaptations require so many theoretical tenets?

I have a number of worries concerning such heavy presentations of the research programme. Of course, some might object to the inclusion of one or two of the theoretical tenets above. Perhaps public policy pursuits are a secondary goal, not a major tenet. Perhaps arguing for massive modularity *a priori* is a niche pursuit. Perhaps. Nevertheless, these positions have been seriously and frequently advocated by prominent evolutionary psychologists, and, as we'll see in the next chapter, such claims are strongly associated with evolutionary psychology, at least in the sceptic's mind.

My concern does not depend on the particular formulation I tabled above. One can formulate evolutionary psychology's purported tenets in alternative ways (e.g. Cosmides tends to stress that the brain functions as a computer, while Bolhuis *et al.* (2011) focus on evolutionary psychology's purported gradualism, the idea that evolutionary change is slow and gradual), and add further theoretical tenets, or less theoretical tenets. And that's precisely my point: what tenets are properly part of the research programme, the activity of proposing and testing adaptationist hypotheses in psychology, and what tenets have been associated with the programme, but which are not strictly part of it in virtue of being independent of the programme's stated goal of proposing and testing adaptationist hypotheses, is rarely addressed.

If we are to treat seriously the idea that evolutionary psychology functions heuristically then it should become immediately apparent that the programme is excessively bundled. In Chapter 1 we saw how leading evolutionary psychologists frequently package evolutionary psychology as something much more than a research programme in the evolutionary behavioural sciences going about its daily business. They present it as a game changer, a scientific revolution, a paradigm shift, the metatheory of the behavioural sciences. Of course, such proponents are using the term 'paradigm' loosely. Just as I'm not trying to draw Lakatos into our discussion by referring to evolutionary psychology as a 'research programme', so too I'm sure that the proponents cited above are not aiming to recruit Kuhn by calling evolutionary psychology a 'paradigm'. Nevertheless, it's clear that in using the term, they're trying to suggest that evolutionary psychology can be much more than operationalising a set of ideas heuristically, that one can take those ideas much further, combining them with other tenets and agendas, presenting this larger package of views as a 'scientific revolution' in psychology and the social sciences—as an all-encompassing view of science and, indeed, the world.

The more ambitious and exclusive one wants to make a research programme, the more exposed one becomes, the greater the danger of ending up arguing for highly controversial theoretical tenets, tenets far removed from the daily activities of the programme. The risk is that one can alienate potential allies in other research programmes and fields, perhaps making them unresponsive or unsympathetic to one's work, making them sceptical of one's work—even hostile.

Evolutionary psychology ought to be packaged as a research programme of discovery, its tenets constructed only according to what its best practice requires, within the broader evolutionary behavioural sciences, rather than packaged as a comprehensive paradigm, as a metatheory of not only the evolutionary behavioural sciences but the behavioural sciences in their entirety.

If evolutionary psychology is packaged excessively, then there is a real risk that its underlying reasonableness and heuristic possibilities will be overshadowed. Naturally, when approaching a research programme, one will first consult the foundational textbooks and foundational edited volumes. If these present and argue for tenets in excess of the practice of the programme, one can get the wrong picture, the wrong end of the stick. And then such a person might be inclined to dismiss the programme wholesale—lock, stock and barrel. Indeed, their distaste might overstep the bounds of ostensible objections.

As shall become clear in the next chapter, I believe this is precisely what has happened, and this sheds some light on why the debate has become so polarised and heated. For now we can just acknowledge that excessive packaging risks serious misunderstanding, whether the risk is realised or not. One might think this is a risk worth taking. One might think not taking this risk is tantamount to endorsing a greasy togetherness, ecumenicalism for the sake of ecumenicalism. But as I shall indicate later in this chapter, this is a risk not worth taking, as to reach its full heuristic potential, evolutionary psychology needs to canvass a wider constituency. For now, my goal is simple: to identify only those theoretical tenets fit for purpose. Or, to put it another way, to distinguish between core tenets and accidental tenets.

What theoretical tenets must we postulate in order to account for and justify evolutionary psychology's daily practice? Put another way, how many of the theoretical tenets commonly associated with evolutionary psychology can we strip away? Quite a few. In order to propose and test adaptationist hypotheses in psychology, you don't need to adopt a belief in massive modularity or argue for massive modularity *a priori*. You don't need to consent to empirical adaptationism. You don't need to investigate public policy implications of your research. You don't need to believe that non-evolutionary work in psychology is rubbish or that adaptationism is the best way to investigate psychological traits or that psychology needs to be unified. To pursue and test adaptationist hypotheses in psychology one requires only three theoretical tenets:

(EP1) Evolutionary theory, with an emphasis on natural selection and adaptation
(EP2) The possibility of psychological adaptations
(EP3) A moderate form of methodological adaptationism

It should be immediately obvious that to formulate and test adaptationist hypotheses in psychology, one does not need to adopt any view on the quality of non-evolutionary work in psychology, nor to adopt the view that psychology needs to be unified, much less that evolutionary psychology will be the metatheory for such unification. Ditto public policy pursuits. Hence, I take it as incontrovertible that we can drop (T6) and (T7) respectively.

It should also be clear that to engage in evolutionary psychology as a heuristic one need not argue massive modularity *a priori*. To see this, suppose it turns out to be the case that there are more domain-general processes in the human mind than the leading champions of evolutionary psychology were originally willing to recognise. Suppose that there is indeed some kind of general intelligence. Does this mean there are no domain-specific processes? No. It is entirely possible that the human mind is organised along both types of process. One can recognise the possibility of a cognitive architecture characterised by domain-general processes and a rich bundle of domain-specific processes. So to pursue psychological adaptationism along domain-specific lines, it is not necessary at adopt a strong stance against domain-general processes. One can be entirely agnostic about the issue or, indeed, happily grant the possibility of domain-general processes while focusing on the possibility of domain-specific processes. Hence, we can drop (T4).

Furthermore, (T5), methodological adaptationism, only needs to be adopted in its moderate form: namely, the claim that it's useful to hypothesise (functionally complex) psychological traits as adaptations. We only need to establish that this is a productive way of doing psychology—which we did in the previous chapter. We do not need to establish that this is the only or correct way of doing psychology or the best way of doing psychology.

A little less obvious is the claim that (T3), empirical adaptationism, can be dropped. Empirical adaptationism and methodological adaptationism often travel together. But this is merely accidental: as Godfrey-Smith (2001) observed, these two types of adaptationism are logically independent of one another.

It's easier to see that one can consent to empirical adaptationism but not methodological adaptationism. One can believe in the power and ubiquity of selection and yet believe it's not useful to treat traits as adaptations, at least not in the manner advocated by evolutionary psychologists. One might hold that the particular methods of evolutionary psychology are useless, or unreliable, or unscientific, or illegitimate. Perhaps even dangerous. Moderate methodological adaptationism is not a given. It can be contested. And it has been.

More difficult to see is declining to consent to empirical adaptationism but consenting to methodological adaptationism. If one is agnostic about empirical adaptationism or rejects empirical adaptationism, why believe that investigating adaptationist hypotheses can be useful? If we live in a world where selection is limited, how can treating functionally complex traits as adaptations be useful? I suggest four scenarios in which one can decline to endorse empirical adaptationism yet consent to methodological adaptationism.

First, if one is agnostic about empirical adaptationism, then one could hold that methodological adaptationism is a useful way of finding out the truth of empirical adaptationism. If there are psychological adaptations, pursuing and testing adaptationist hypotheses is a useful way of finding this out. Second, even if one rejects empirical adaptationism, one could still acknowledge that it can lead to new discoveries, and can stimulate new lines of research, even if these discoveries are ultimately explained by non-selectionist factors. Third, even if one rejects empirical adaptationism, investigating adaptationist hypotheses first might be a good starting point. Godfrey-Smith (2001: 342) cites an observation by Kim Sterelny that 'methodological adaptationism might be particularly useful if non-selective factors like developmental and genetic constraints are elusive and hard to discover'. When adopting optimality as a kind of null hypothesis, deviation 'provides evidence that other factors are at work, and perhaps the nature of the deviation will give clues about where to look next' (*ibid.*) Fourth, an admittedly unlikely but nevertheless possible scenario, one could reject empirical adaptationism but hold that methodological adaptationism is a useful way of establishing the falsity of empirical adaptationism. The continual failure of adaptationist hypotheses in the long run would discredit empirical adaptationism.

We shouldn't confuse the daily practice of the programme with the ambitions of those at the summit of the programme. I cannot look into the hearts of individual evolutionary psychologists, but I see little evidence that rank-and-file practitioners are beholden to evolutionary psychology as a paradigm. At the very least, many of the theoretical tenets explicated in Chapter 1 are completely absent in the research papers of evolutionary psychologists. Such research expresses not an all-encompassing view of science and the world, but a heuristic programme, a programme of discovery.

Furthermore, there's evidence that practitioners behave more as members of one research programme among several in the evolutionary behavioural sciences than as members of an all-encompassing paradigm.

Here, I draw upon recent research by Machery and Cohen (2012). They are unique in the philosophy literature on evolutionary psychology in that they used quantitative citation analysis to judge four commonly held hypotheses about the evolutionary behavioural sciences. One of the hypotheses they consider is whether the evolutionary behavioural sciences divide into several distinct competing paradigms. Machery and Cohen (2012: 184) distinguish between 'research traditions' and 'paradigms' as follows:

> First, research traditions within a given scientific field (e.g. psychology of vision) share various commitments—they might agree on what the explananda are, they might share some methodological principles, or they might concur on what the characteristics of

> successful explanations are—while paradigms have little in common. Second, two hypotheses formulated within two distinct research traditions might both be correct, while the hypotheses formulated within two distinct paradigms tend to be incompatible.

If the hypothesis that evolutionary psychology and human behavioural ecology are competing paradigms with little in common is true, then evolutionary psychology and human behavioural ecology papers should display noticeably different citation patterns. Machery and Cohen performed a quantitative citation analysis on articles published in *Evolution & Human Behavior* in the period January 2000 to December 2002. As one would expect, the citation patterns were not identical, there were striking similarities: 'evolutionary psychologists and human behavioral scientists appeal to evolutionary biology and non-evolutionary behavioral sciences in the same proportion—that is, they are influenced by these two fields to the same extent' (*ibid.*: 216). Because these similarities are greater than one would expect of competing paradigms, Machery and Cohen concluded that evolutionary psychology and human behavioural ecology form two distinct research traditions, not two competing paradigms.

Other evidence also suggests that evolutionary psychology is practised as one of multiple research programmes or traditions within the evolutionary behavioural sciences, rather than practised as a rival paradigm for doing evolutionary behavioural science. For example, researchers from different traditions and programmes in the evolutionary behavioural sciences attend common meetings and conferences, such as the meetings of the Human Behavior and Evolution Society (Machery and Barrett, 2006; Brown *et al.*, 2011; Machery and Cohen, 2012). Furthermore, recent edited volumes of evolutionary psychology work include contributions from behavioural ecologists (e.g. Barkow, 2006; Swami, 2011).

So what we've done is to distinguish between evolutionary psychology as a research programme and evolutionary psychology as a paradigm. If one is uncomfortable with those labels, alternative labels are available. For example, we can apply Godfrey-Smith's distinction between science and philosophy of nature. Whereas science is a set of ideas and tools for investigating the world, a philosophy of nature is 'taking science as developed by scientists, and working out what its real message is, especially for larger questions about our place in nature' (Godfrey-Smith, 2009: 3). Prominent evolutionary psychologists clearly go beyond the science and make frequent philosophy of nature pronouncements.

Recall the Duntley and Buss (2008) quote in Chapter 1, where Duntley and Buss not only claim that evolutionary psychology unites the social sciences, but that it also 'unites humans with all other species, revealing our place in the grand scheme of the natural world' (2008: 31). Yet, to adopt the science of evolutionary psychology, to use its ideas and methods to investigate the world, we don't need to adopt a philosophy of nature—and if we do, we're free to adopt an alternative philosophy of nature to that subscribed by prominent evolutionary psychologists.

Further alternative labels are available: one can distinguish between a bottom-up evolutionary psychology and a top-down evolutionary psychology, between a streamlined evolutionary psychology and an inflated evolutionary psychology, between a modest evolutionary psychology and a vaulting ambition evolutionary psychology ('vaulting ambition', a phrase from Kitcher, 1985). Pick whichever label you prefer.

Streamlining the framework of evolutionary psychology to better reflect this reality offers multiple rewards. First, it finally shuts the lid on a range of objections that keep springing up like a jack-in-the box, objections that fail to target, let alone undermine, the core of the programme, being based on periphery tenets, or overly dramatic interpretations of those tenets. For example, evolutionary psychology is frequently derided as being *hyperadaptationist* (Gould, 2000), as believing that every splinter, every speck, of the mind is caught within the orbit of selection. Nonsense. Investigating a range of psychological traits to see if any have been selected for does not mean one believes every psychological trait is an adaptation. A streamlined evolutionary psychology is not committed *a priori* to a viewpoint on the number of psychological adaptations that exist. It's entirely an empirical matter. It might be fewer than prominent evolutionary psychologists believe, or as many as they believe, or even more—from just a handful of functionally specialised psychological adaptations, to less than a hundred, to 'hundreds or thousands' (Tooby and Cosmides, 1995: xii), to even tens of thousands. Or, indeed, none.

Second, a bottom-up presentation reduces the space for misunderstandings to arise, for misunderstandings to be crystallised and recycled, and so lessen the need for corrections, and continual corrections, to be made. Those who present evolutionary psychology top-down tend to make a number of global claims. They claim that our 'modern skulls house a stone age mind', and make claims about 'human nature' and 'the psychic unity of humankind' (Cosmides and Tooby, 1997a). Such global claims are easily misunderstood, and have been commonly misunderstood to mean we are behaviourally inflexible, behaviourally invariable and behaviourally poorly adapted to today's environment (as we saw in Chapter 2). However, when we present evolutionary psychology as a bottom-up enterprise, we're less inclined to make global pronouncements, less inclined to make philosophy of nature statements and so less likely to be misunderstood. A bottom-up evolutionary psychology simply proposes and tests a range of adaptationist hypotheses. Such research can also be presented as explanations. But the evolutionary psychologist need not go beyond this. If she wishes to make claims about human nature and the mind generally, she can do so, in a separate capacity, wearing a different hat to the one she wears proposing and testing individual adaptationist hypotheses.

Third, a bottom-up presentation of evolutionary psychology focuses exclusively on daily practice, not drawing attention away from daily practice, and so better serves the programme. With the clutter stripped away, the basic picture and the possibilities the programme should be clearer and easier to grasp—to both the practitioner and the spectator.

Fourth, streamlining removes many barriers and obstacles to entry. It becomes much clearer that pursuing adaptationist hypotheses in psychology need not commit us to particular philosophies of nature. The use of the tool is consistent with a variety of views, whether one emphasises selection or whether one emphasises drift, whether one believes in free will or not, whether one believes the sciences should be unified or not.

Fifth, streamlining the presentation of evolutionary psychology reduces friction with other programmes or traditions and, accordingly, allows us to better grasp their underlying compatibility. Much of the diversity in the evolutionary behavioural sciences can be accounted for as differences in focus and differences in emphasis. As Machery and Cohen (2012: 184) note, 'hypotheses developed in each tradition are often compatible in that they focus on different phenomena or on different aspects of the same phenomena'. So it can only be a benefit if evolutionary psychology is not unnecessarily positioned in conflict with its immediate adjacent disciplinary neighbours.

For example, recall the previous chapter, where we discussed evolutionary psychology's by-product account of religion. Does this exhaust an evolutionary understanding of religion? Clearly not: there is plenty of space for other types of evolutionary explanation too, and such explanations are unlikely to find themselves in conflict with the evolutionary psychology accounts as they're focusing on different aspects of the phenomenon of religion. Dawkins recognises this possibility: 'Even though conventional Darwinian selection of genes might have favoured psychological predispositions that produce religion as a by-product, it is unlikely to have shaped the details ... if we are going to apply some form of selection theory to those details, we should look not to genes but to their cultural equivalents' (2006: 190).

Furthermore, while human behavioural ecology focuses on the reproductive consequences of behaviour, evolutionary psychology focuses on the mechanisms that generate behaviour. Whereas a strongly packaged evolutionary psychology can express, and has expressed, scepticism about 'counting babies', stressing that the advent of reliable contraception in modern societies problematises such approaches, a streamlined evolutionary psychology doesn't comment on other research programmes.

Our different temperaments, sensitivities, schooling, skills and disciplinary pedigrees will naturally incline us towards one tradition or programme over others, but just because we elect to work within one tradition or programme does not mean we must believe that all researchers should work in the same programme. In the absence of a unifying framework, skirmishes and flashpoints between traditions are possible, and have indeed taken place, but we need not overplay this. We need not declare for one programme against another. We can be greedy—we can try to eat the entire cake. Indeed, in order to satisfy our hunger for complete explanations, we'll likely need to eat the entire cake.

The literature is increasingly recognising the underlying compatibility and possibilities for convergence. Recognising the diversity of approaches in the evolutionary behavioural sciences, Laland and Brown (2002: 296) observe,

> Irrespective of the methodological differences among the practitioners, there is little that is conflicting or incompatible about these findings. In fact, each investigation reinforces the others, collectively building up a panoramic view of the topic at hand that spans genetic to sociocultural levels of analysis and transects distant continents. Here is an advertisement for pluralism in evolutionary perspective. There is no reason for researchers to restrict themselves to a single research technique when, by and large, the different methodologies are highly complementary.

Indeed, integrated studies, studies that combine the conceptual and methodological resources of two or more traditions, are becoming an important feature of the contemporary literature (Sear *et al.*, 2007).

Two final clarifications. First, I am not claiming that the expunged or divorced tenets are not worth discussing. They absolutely are. But they are independent of evolutionary psychology practice. Second, I am not claiming that the divorced tenets are false. I have no quarrel with empirical adaptationism. And though I see no pressing need for unification of psychology, and indeed think we have good reasons to favour disunity in psychology and the behavioural sciences, I'm not especially animated about the issue of whether the behavioural sciences need a metatheory and not interested in getting into a drawn-out conceptual fight about it. If individual evolutionary psychologists want to discuss those issues and advocate particular positions, fine—but doing so should be sharply distinguished from the research programme, not casually or blindly mixed in with the research programme. Proposing and testing hypotheses and pursuing, say, a reforming agenda in the social sciences are independent projects. If this is not made clear, if they're packaged together, and packaged with other contentious tenets, all under one banner, this threatens to overshadow the research programme. Keep different agendas separate.

An analogy will be instructive. The first draft of a paper will contain a multitude of points. Often, some of these will obscure the really valuable points. The paper, therefore, requires editing; it requires streamlining, so that its proper focus is brought fully to light. Sometimes this will require the removal of interesting material, but such must be done. And so it is with evolutionary psychology's purported tenets. The trimming is overdue.

4.2 The Challenge of Adaptationist Explanation Revisited

So framing evolutionary psychology as a heuristic project, rather than as a paradigm, happily liberates the programme from much of the criticism aimed against it. Importantly, it also deflects criticism based on the incompleteness of its explanations—criticism that is the source of much unease and mistrust about the programme. Recall from Chapter 2 that Richardson (2007) and others have adopted the hardest of positions against evolutionary psychology. But in light of our reframing of evolutionary psychology as a heuristic project, how sustainable is that position?

The daily practice of evolutionary psychologists is to formulate hypotheses and test them. These can be, and often are, presented as explanations. The lesson from Richardson, and the long train of sceptical thought on which he rides, is well taken: these are not complete explanations as they lack a great deal of evidential particulars.

Notice, it's not necessary to formulate evolutionary psychology research as explanations—one could limit oneself to proposing a range of testable hypotheses: *if* trait *T* is an adaptation for *X*, trait *T* should have configuration *C*, and so we should find phenomenon *P*. One need not then make the further claim that trait *T is* an adaptation for *X*. In other words, strictly speaking using adaptationism as a heuristic does not commit us to using adaptationism as an explanation. The evidential requirements needed to vindicate or complete an adaptationist explanation, the set of statements giving information or details about the relevant ancestral populations and ancestral environments such as the trait variation present in ancestral populations and information on the demographic structures of ancestral populations, only become active when we present such research as explanation. Nevertheless, Richardson would point out that evolutionary psychologists routinely do take this further step: 'This is the root problem that evolutionary psychologists face: they claim that human psychological traits are adaptations and thus need to *explain* them *as* adaptations' (2007: 98, original emphasis).

Sceptics frequently lampoon what they see as the naive confidence of evolutionary psychologists when making the move from hypothesis to explanation. Sceptics are beholden to a picture of evolutionary psychologists testing a limited number of hypotheses on a limited number of subjects, and then declaring full-blown claims about human nature and working up great public policy recommendations. Kitcher (2009: 91–2) captures the impression well:

> I would be less concerned to hold evolutionary studies of human psychology and behavior to high standards, if we lived in a world in which the findings of such studies were soberly presented as preliminary, in which the difficulties of devising adequate evolutionary models of human psychology and behavior were clearly understood and acknowledged, in which 'results' were not trumpeted to the general public as exiting new discoveries, firmly established.

If evolutionary psychologists really do believe that their hypotheses stand as complete, established explanations, and market them as such, then this should, indeed, be highlighted and challenged. Sceptics shouldn't stand for it.

Happily, they don't need to. The lesson was taken and digested long ago—if, indeed, the lesson was ever really needed. For example, in a review of Jerry Fodor and Massimo Piatelli-Palmarini's *What Darwin Got Wrong*, Barrett (2011: 76) observes that the authors provide

> the kind of cautionary note that seems all the rage amongst philosophers writing about 'Darwinism' at the moment: a scolding to those of us who are perceived as running amok in the playground of adaptationist explanation (e.g., Buller, 2005; Dupré, 2001; Richardson, 2007) ... I happen to find most of these scoldings a bit tiresome—how often do we have to be reminded that all traits of organisms aren't adaptations, or that hypotheses should be evaluated against alternatives?

Absolutely. Perhaps some evolutionary psychologists need to be more modest in what conclusions they draw from the successful corroboration of their hypotheses. Perhaps some evolutionary psychologists need to be more alert to alternative explanations. But this is true of every discipline. Name me a discipline where this isn't true. A shiny penny for every research programme about which this could not be said. The warning not to rush to judgment or jump to conclusions might sound reasonable enough, but the continuous repetition of this point, the continuous *recycling* of this point, is subversive: this is how legitimate viewpoints can be marginalised and patronised away. How many times do educated professionals need to be told that while 'evolutionary analyses may generate clues as to the mechanisms of human cognition, these are best regarded as hypotheses, not established explanations, that need to be tested empirically' (Bolhuis *et al.*, 2011: 4)?

A sceptic might point to the avalanche of public news about evolutionary psychology findings, how sensationalist they are given their incompleteness with respect to evidential particulars. A sceptic might hold that evolutionary psychologists really do need to be schooled in how to interpret their findings in a way that acknowledges the limitations of what successful experimental outcomes demonstrate. But, in response, we should counter that evolutionary psychologists can't be held responsible for how their research enters the public domain in popular science newspaper articles and then disseminated. Science reporting is rife with sensationalist interpretations of quite limited studies. We'd take a dim view of someone who thought quantum physicists need to be re-schooled in the basics of science just because their work is occasionally sensationalised in the popular media.

I want to leave that point to one side now—I believe it's been sufficiently made. The deeper point I wish to raise is this: the charge that evolutionary psychology explanations—when their research is presented as explanation—are incomplete is well taken but, crucially, we should not take from this that their research is therefore methodologically sterile, inert, that all that they're producing is 'just so' stories. Indeed, such an inference is deductively invalid. Even if the evidence required to fully vindicate adaptationist explanations is lacking, this fails to undermine the notion that much is to be gained through adaptationist hypothesising in psychology. Hence, neither the current paucity of knowledge of the past nor pessimism about prospects of gaining such knowledge in anyway destabilises the legitimacy and utility of evolutionary psychology. A US university textbook on the biology of behaviour makes the eminently sensible observation that lack of certainty concerning our evolutionary past does not frustrate the utility of the evolutionary perspective in psychology: 'Although the events that led to the evolution of the human species can never be determined with certainty, thinking of the environmental pressures that likely led to the evolution of our brains and behavior often leads to important biopsychological insights' (Pinel, 2008: 3).

None of the considerations Richardson tables have any bearing on whether evolutionary psychology is useful or not, on whether it's merely speculation or not. Indeed, one could argue for further explanatory and evidential requirements, or challenge some of the items on Brandon's list, but as evolutionary psychology primarily operates in the context of discovery, not the context of justification, the point is rather moot.

Indeed, Richardson inadvertently hints at this. Usually, Richardson dismisses evolutionary psychology wholesale, but he occasionally softens his line a little. For example, we read,

> I do not generally claim that these evolutionary hypotheses concerning human psychology are false. I do think that some do not warrant serious consideration, for lack of evidence. For all intents and purposes, we can dismiss them. Some may eventually yield to empirical discoveries. If this is so, then so be it (Richardson, 2007: 38).

Clearly, hypotheses yielding empirical discoveries are not 'idle', are not merely 'speculation' and we shouldn't respond to such discoveries with the attitude 'then so be it'.

So although detailed ancestral population information, knowledge of variation and drift in ancestral populations, is certainly highly desirable, evolutionary psychology can nevertheless successfully progress without it. And, crucially, the progression of evolutionary psychology can help orchestrate activity that might bring about more complete explanations in the long run.

4.3 Evolutionary Psychology in the Evolutionary and Behavioural Sciences

I don't wish to shirk from the spirit of this complaint, for in rising to the challenge of adaptationist explanation we better locate evolutionary psychology's role in the evolutionary and behavioural sciences. I argue that it's for the evolutionary and behavioural sciences collectively to rise to the challenge of adaptationist explanation.

In the previous chapter, I postulated the following three heuristic functions of evolutionary psychology:

H1 Hypothesising unknown design features of extant psychological mechanisms, leading to novel predictions of psychological phenomena
H2 Hypothesising unknown psychological mechanisms, leading to novel predictions of psychological phenomena
H3 Hypothesising traits as by-products of psychological mechanisms, leading to novel predictions of psychological phenomena

Now I identify and argue for another function of evolutionary psychology:

H4 Stimulating multidisciplinary research activities, leading to more sophisticated hypotheses and more complete ultimate and proximate explanations

An evolutionary psychology hypothesis postulates an adaptation, or a by-product from a set of adaptations, and a set of observable consequences that should obtain. When successfully corroborated, this produces a new line of evidence. But a new line of evidence is insufficient to establish a complete explanation. It's for the evolutionary and behavioural sciences collectively, not evolutionary psychology alone, to establish the required evidential particulars.

Richardson and other sceptics fail to pick up on this. We must, however—and we shall.

In order to understand properly a psychological adaptation, we need not only a complete ultimate explanation, as highlighted by Richardson and others, but also developmental and neurological details—in other words, we need complete evolutionary (ultimate) explanations and complete proximate explanations. Or, to use Tinbergen's (1963) fourfold distinction, we need information on the trait's function, phylogeny, ontogeny and mechanism. Whether one presses for further evolutionary details, or further proximate details, or both, it should be obvious that establishing such details calls for multiple lines of evidence—from multiple disciplines.

Establishing ultimate and proximate details requires the tools, techniques, methods and knowledge from the full range of the evolutionary and behavioural sciences. It requires not only evolutionary psychology, but also archaeology, evolutionary anthropology, genetics, behavioural genetics, behavioural ecology, evolutionary biology, developmental biology, developmental psychology, comparative psychology, cognitive psychology and evolutionary cognitive neuroscience.

Evolutionary psychologists cannot be Jacks of all trades. Kitcher (2009: 81–2) claims,

> Evolutionary psychologists must *themselves* do work that is as precise in its modeling, as painstaking in its accumulation of data, and as sensitive to alternative hypotheses as that undertaken in a large number of admirable evolutionary studies. Yet even in the most celebrated examples, the work of Leda Cosmides and John Tooby on detection of cheating, for instance, the modeling is pitifully crude in comparison to that found in mainstream evolutionary theorizing, the knowledge of environmental details is rudimentary and the underlying genetics simply absent (original emphasis).

Coyne (2009: 230) express a similar sentiment:

> We should be deeply suspicious of speculations that come unaccompanied by hard evidence. My own view is that conclusions about the evolution of human behavior should be based on research at least as rigorous as that used in studying nonhuman animals. And if you read the animal-behavior journals, you'll see that this requirement sets the bar pretty high, so that many assertions about evolutionary psychology sink without trace.

Why must evolutionary psychologists themselves do all this work? Yes, evolutionary psychologists should be aware of alternative hypotheses, but why should evolutionary psychologists alone accumulate the full portfolio of data required of a complete adaptation explanation, data that requires a multitude of methods and disciplinary skills? Why should they alone establish the full portfolio of developmental knowledge of how genes and environments build specific adaptations? Are critics seriously expecting evolutionary *psychologists* alone to do all this work?

Evolutionary psychologists cannot be expected to engage in genetic and developmental research. They cannot be expected to sequence genomes, to hunt for molecules, to dig around archaeological sites, to enter caves deep and ancient. They cannot be expected to perform paleontology, paleobiology, arrange and partake in expeditions to remote hunter–gatherer societies nor to train part time for PhDs in neuroscience.

As we saw in Chapter 3, evolutionary psychologists tend to make psychological predictions and perform psychological experiments, exactly as is to be expected from their disciplinary pedigree. Exactly what one would expect of *psychologists*. They primarily test predictions using self-reports, archive data and behavioural studies. That's their speciality. That's what they're trained to do.

Evolutionary psychology's role in the evolutionary and behavioural sciences is discovery. Its methodology is crafted for that very role. But evolutionary psychology's tools alone cannot complete the task of furnishing complete explanations. Evolutionary psychology can *contribute* to the portfolio of evidential details in virtue of its successful novel predictions but its *modus operandus* is in the context of discovery. So evolutionary psychology alone cannot do much of what the sceptics demand of it, but the bigger point is that it can stimulate research by other research programmes into gathering such evidence. It is for a wider constituency—the evolutionary and behavioural sciences collectively—to unearth the evidence required to authenticate and justify evolutionary psychology hypotheses.

Schmitt and Pilcher (2004) argue that to support a psychological adaptation claim, evidence should be presented from eight disciplinary lines of support: psychological evidence (e.g. surveys and behavioural tests), medical evidence (e.g. fertility and mental health studies), genetic evidence (behavioural genetics, population genetics, molecular genetics), physiological evidence (e.g. neuroanatomical structures), phylogenetic evidence (comparative psychology, primatology, physical anthropology), hunter–gatherer evidence (from anthropology and also human behavioural ecology), cross-cultural evidence (cross-cultural behavioural patterns and ecology-dependent variability) and theoretical evidence (e.g. evolutionary game theory simulations, computer modelling and cost–benefit analyses). Of course, one or two of these categories might be questioned. For example, one might argue that contemporary hunter–gatherer studies are of limited relevance, while another might argue that contemporary hunter–gatherer studies offer important evidence of ancestral activity patterns. Regardless, it should be incontrovertible that vindicating an adaptation claim requires resources from a number of evidence categories.

Obviously, evolutionary psychologists directly contribute to the first category, the category of psychological evidence. As the postulation of a psychological adaptation raises empirical questions and issues in cognate and adjacent fields, there is no reason why evolutionary psychologists can't work in tandem with these disciplinary fields. Indeed, this is already happening, albeit slowly, albeit in a piecemeal fashion.

When postulating a psychological adaptation, perhaps the most immediate question is, how cross-cultural is the trait? What ranges of expression can be found cross-culturally? Is there predictable variability of expression based on local environmental (social and ecological) and developmental conditions? Cross-cultural studies are where notable collaborations have arisen. For example, during the last few years, the International Sexuality Description Project, a collaboration involving over 100 biological, behavioural and social scientists, has conducted several series of tests to examine whether evolutionary psychology predictions obtain cross-culturally. In one series of tests, Schmitt *et al.* (2003) tested for an evolutionary psychology hypothesis concerning sex differences in the desire for sexual variety, surveying 16,228 people via anonymous self-reports in 52 nations across the 10 major world regions—North America, South America, Western Europe, Eastern Europe, Southern Europe, the Middle East, Africa, Oceania, South Asia and East Asia. This remains one of the largest cross-cultural investigations undertaken on whether the sexes differ in the desire for sexual variety. The 118 biological, behavioural and social scientists who co-authored the paper, and who collected and analysed the test results, found that the predicted sex differences obtain. Across cultures, men tend to not only have a greater desire for a variety of sexual partners than women, but they also tend to require less time to pass before agreeing to sexual intercourse. Furthermore, across cultures, men tend to more actively seek short-term relationships than women.

Evolutionary psychologists are also collaborating with neuroscientists. Platek and Shackelford (2009) is a collection of several papers in the emerging field of evolutionary cognitive neuroscience, which combines evolutionary psychology with cognitive neuroscience. Takahashi and Okubo (2009) investigated sex differences in the neural correlates of jealousy using functional neuroimaging techniques. They observed sex differences in brain activations in response to sexual and emotional infidelity scenarios, 'supporting the view that men and women have different neurocognitive systems to process a partner's sexual and emotional infidelity' (*ibid.*: 211)

Takahashi and Okubo (2009) noted that, for ethical reasons, it's not possible to investigate brain activity in response to actual instances of infidelity. Instead, as per common practice, the researchers investigated brain activity in response to hypothetical infidelity scenarios. Indeed, as Kitcher and Vickers (2003) note, certain types of experiments that could make significant progress in evolutionary studies of psychology are forbidden as unethical. We cannot genetically or neurologically manipulate humans, nor can we deprive humans of certain developmental contexts. We are handicapped in some respects when studying humans. But we need not overplay this. First, this is not a problem unique to evolutionary psychology: psychologists of all stripes are often ethically bound to postulate hypothetical scenarios on their human test subjects, rather than present actual ones. Second, studying non-human animals also has its handicaps—one cannot, after all, ask a primate his or her feelings or preferences on various topics of interest. Third, there are also 'natural experiments', so to speak, that can be appealed to and analysed—such as those unfortunate enough to suffer a genetic-based disability, those unfortunate enough to suffer accidents, twin studies and studies of feral children.

Evolutionary psychologists can, and sometimes do, think across species. For example, Kramer *et al.* (2011) discovered that untrained humans can accurately judge aspects of chimpanzee personality from pictures of static, non-expressive chimpanzee faces alone. On the strength of this result and other studies, they identified the possibility of a shared signal system between the two species: 'We hypothesize a shared signal system for personality from the face in humans and chimpanzees; that is, on the basis of their shared evolutionary past, chimpanzees and humans share aspects of a system for communicating behavioral biases to conspecifics' (2011: 183). Kramer and Ward (2012) replicated and extended those findings using a new group of chimpanzees.

Comparative studies can also be used to see precisely how a well-designed adaptive solution should operate. For example, from comparative studies we have a good idea of what an adaptation for infanticide, of killing non-relatives within a group, looks like. When taking over harems of females, male gray langurs tend to kill all infants under six months old that are not their own. If one were to design an adaptation for infanticide, one could do no better than to design behavioural strategies exhibited by male gray langurs. Do human stepfathers act like this? Clearly not. As we saw in Chapter 1, according to Daly and Wilson (1980), stepfathers represent the most significant risk factor for child abuse. Nevertheless, stepfathers simply do not systemically behave in the way predicted by a direct adaptation for child abuse or infanticide. Such behaviour has the hallmarks of being a by-product. So such a comparative study can help vindicate an explanation of stepchild abuse as being a by-product of a psychological adaptation, rather than directly as an adaptation of its own.

It should be no surprise that evolutionary psychologists rarely use phylogenetic tools. Phylogenetic analysis is a skill set. It takes training to be proficient in such a skill, and time to become suitably knowledgeable in it. Perhaps evolutionary psychologists could become more aware of relevant phylogenetic analyses, and perhaps a few might even opt to train to use this tool. But this falls short of requiring evolutionary psychologists themselves to generate phylogenetic information concerning primitive and derived characteristics. Instead, phylogenetic specialists can work in tandem with evolutionary psychologists.

And it should be no surprise that evolutionary psychologists rarely cite genetic research. Kitcher and Vickers (2003: 337) misread this as a tactical ploy: 'the latest Darwinizers have learned from the demise of old-style pop sociobiology: Be cagey about genetic hypothesizing!' Caginess? That's not quite right. Evolutionary psychologists focus on developmentally reliable adaptations, not on genes. And again, perhaps geneticists and developmental biologists knowledgeable of evolutionary psychology work, or working in tandem with evolutionary psychologists, can make progress in this direction.

Theoretical evidence can take the form of evolutionary game theory simulations. Evolutionary game theory provides conceptual tools to establish which behavioural strategies can spread and stabilise in a population over time. Evolutionary game theory models and studies the interactions of boundedly rational, self-interested agents playing games that model decision problems arising in social contexts. The change of strategies, which is a process of adaptation, is one of the features that distinguish evolutionary game theory from classic game theory. The 'evolution' in evolutionary game theory means nothing more than changes in belief over time. Hence, changes of strategy or adaptations can be interpreted to represent either biological selection or cultural learning.

Evolutionary game theory can be taken to cover a wide and diverse set of games. Games encapsulate important aspects of different bargaining problems that underlie various behavioural phenomena, allowing us to think about evolution where the advantage of behaving in a certain way depends on the behaviour of others. Clearly, in order to get a firm grip on this, a lot of particulars present in a given social phenomenon need to be stripped away; this is precisely what the evolutionary game theoretic approach enables social scientists and philosophers to do.

Four of the most common games, which are taken to represent different aspects of moral phenomena, are Divide the Cake (also known as 'Divide the Dollar'; represents fairness), the Prisoner's Dilemma (represents co-operation), the Stag Hunt (represents trust) and the Ultimatum Game (represents retribution).

An example of such evolutionary game theoretic work is Skyrms (1996), a model of the evolution of fairness. Two individuals, presumably cake lovers, are fortunate enough to be offered a cake to divide between themselves. How should they divide it? Quite naturally, all things being equal, we would hold that a 50–50 split between the two is the only fair or just solution to this particular division problem. This seems obvious. But why we do understand distributive justice like this? There are an infinite number of ways of carving up the cake. Why don't we egg on our lucky cake-eaters to adopt a 70–30, 80–20 or 90–10 split, and so on? Rational deliberation does not settle the matter, so why is the social norm of fairness so widespread?

Skyrms studied the evolutionary dynamics of the Divide the Cake game in order to shed light on why the social norm of fairness is universal. In Divide the Cake, two players are offered a cake and are required to demand a certain portion of it. If their requests add up to less than the full cake, both players get what they want. However, if their requests are in excess of the full amount, neither player receives anything. For reasons of simplicity, three strategies can be put on the table: (S1) always demand half of the cake (fair); (S2) always demand two-thirds of the cake (greedy); (S3) always demand one-third of the cake (modest). The fair split is an evolutionarily stable strategy (ESS). As defined by Maynard Smith and Price (1973), an ESS is any strategy that, once every agent in a population has adopted it, cannot be displayed by any alternative behaviour strategies that try to invade the population. So, once every agent in a population requests the 50–50 split, this state will persist over time.

What is common to all evolutionary game theoretic simulations is that a number of agents in a population, typically a large number, are matched in order to play a round of a particular game (the way agents are matched differs between games). These agents receive a payoff according to the rules of the game in question. They then change their behavioural strategies according to a given dynamic process. The replicator dynamics is one such dynamic process, according to which strategies that do better than average increase within the population over time, and those that do worse decrease. The agents are then matched again in order to play another round of the game in question. This process is continued until the population converges to a stable distribution of strategies.

Evolutionary game theory is primarily an instance of what D'Arms *et al.* (1998) call 'evolutionary generalism', which seeks to model and explore behavioural strategies and their interactions and payoffs within populations; it's not within the explanatory ambitions of evolutionary generalist approaches to identify the proximate psychological mechanisms underlying those behaviours. In other words, evolutionary game theorists do not open the 'black box' (Ernst, 2005). Evolutionary psychology is the 'evolutionary particularist' enterprise that seeks precisely to do that.

Proposing novel hypotheses occurs all the time in other parts of the evolutionary sciences without the slightest hint of controversy. As Dennett (2011: 482) notes,

> It is worth noting that evolutionary biologists confidently hypothesize historical events—horizontal gene transfers, for instance, that occurred billions of years ago, give or take a few hundred million years, or speciations and migrations that must have occurred at some point, ill-defined in space, time and causation—without fear of being chastised for indulging in Just So Stories. It is pretty much only hypotheses about human evolution that are held to a higher—conveniently unattainable—standard of evidence, by the critics of sociobiology or evolutionary psychology. Yes, there are egregious cases of hypotheses being defended solely on grounds of their plausibility, given the few facts available, but they shade into entirely reasonable cases—across biology, so far as I can see—with no clear boundaries. Much of the progress in evolutionary biology consists in the confirmation or disconfirmation of bold hypotheses that started out as plausible guesses.

Many hypotheses in the evolutionary sciences are born naked, lacking great swathes of information, and only gradually do they become clothed in detail. They are not dismissed as mere speculation. No-one demands that they immediately enter the world fully formed. They are seen for what they are: the first step in perhaps a new, promising, and potentially long and fruitful, line of research, one that can trigger multiple disciplinary engagements, perhaps leading to modification of the original hypothesis or explanation, perhaps leading to abandonment, perhaps someday to acceptance and consensus.

Laland and Brown (2011b) discuss the co-evolution of dairy farming and lactose tolerance, which we touched upon in Chapter 2. Simoons (1969, 1970) originally hypothesised that dairy farming created selection pressures that led to genes facilitating lactose tolerance becoming common in dairy farming societies. What is striking is how interdisciplinary subsequent research in this area was. Laland and Brown note that a variety of researchers—anthropologists, archaeologists, geneticists, and so on—engaged with Simoons' hypothesis in their various ways. Geneticists identified a gene responsible for lactose tolerance, and statistical geneticists confirmed that it shows strong signs of recent selection. Phylogenetic methods and statistical methods were deployed, both to rule out alternative explanations and to further support the dairy farming hypothesis. Furthermore, on the strength of the hypothesis, novel predictions were made. Laland and Brown point out that Durham (1991) predicted that populations with traditions for consuming fermented milk products but not fresh milk should have intermediate levels of lactose tolerance, as fermented milk products have lower levels of lactose than fresh milk. The novel prediction was subsequently empirically supported.

Forty years of multidisciplinary research has produced an impressive portfolio of evidence vouching for the co-evolution of dairy farming and lactose tolerance hypothesis. In principle, there is no reason why successful evolutionary psychology hypotheses, those that produce successful novel predictions, making genuine contributions to knowledge, and that survive refutation attempts, cannot stimulate cognate and adjacent fields to pursue further research in directions evolutionary psychologists alone cannot practically reach, to establish better, stronger portfolios of evidence for postulated psychological adaptations.

Such multidisciplinary co-operation takes time. It also requires an awareness and knowledge of the status of purported psychological adaptations, as well as an ecumenical spirit between the various programmes and disciplines in the evolutionary and behavioural sciences. There are, however, stumbling blocks to cross-disciplinary understanding and ecumenicalism. First, researchers in adjacent disciplines might only have a rudimentary understanding of the state of play in evolutionary psychology, only being aware of a handful of purported psychological adaptations, and will often have little idea as to what dossier of evidence is available for each purported psychological adaptation. Second, the evolutionary psychology debate is highly polarised. While this polarisation might be less pronounced on the ground than it is in the literature, it does, nevertheless, have a detrimental effect.

I'll tackle the second problem in the next chapter. Here, I identify and discuss a forthcoming development that has every chance of transforming the first problem. Balachandran and Glass (2012) report on the forthcoming launch of 'PsychTable', a taxonomy aiming to codify the number of psychological adaptations that have been hypothesised in the literature, and to evaluate the strength of evidence for each purported psychological adaptation by aggregating both the supporting and negative evidence available. The taxonomy leverages Schmitt and Pilcher's (2004) identification of eight categories of evidence needed to vindicate a psychological adaptation claim. Against these eight lines of evidence, the PsychTable can identify both the evidentiary breadth and the evidentiary depth of each purported psychological adaptation.

Evidentiary breadth pertains to the range of evidence an adaptation claim enjoys. According to Balachandran and Glass, citing Schmitt and Pilcher, at minimum, a purported psychological adaptation must have evidence falling into one of the eight evidence categories. A 'moderate' level of evidential breadth is held to be an evidence portfolio spanning two or three of the evidence categories. An 'extensive' level of evidential breadth is when four or five evidence categories can be cited. Finally, one can judge the evidential breadth of a purported psychological adaptation to be 'exemplary' when six of more evidence categories can be cited.

Evidentiary depth pertains to the quality of evidence an adaptation claim enjoys. According to Balachandran and Glass, again citing Schmitt and Pilcher, single studies with poor methodological control or unrepresentative sampling represent minimum evidentiary depth. At least two studies with good levels of methodological control and good sampling quality represent a 'moderate' level of evidentiary depth. Several studies with high levels of methodological control and high sampling quality represent 'extensive' evidentiary depth. Finally, multiple studies with the highest levels of control and sampling represent an 'exemplary' level of evidentiary depth in favour of a purported psychological adaptation.

The PsychTable will aggregate the citations using an algorithm to evaluate the strength of support for each purported psychological adaptation, automatically assigning an overall evaluative score, as well as generating a table indicating currently available evidence. As you might expect, it promises to offer a rich harvest of gains desperately needed in the literature. First, by codifying purported psychological adaptations and evaluating their breadth and depth of evidence, it will become immediately apparent to researchers, both within evolutionary psychology and in adjacent disciplines, where further research is needed. As Balachandran and Glass (2012: 313–14) note,

> PsychTable will allow researchers and contributors to aggregate studies and evidence from across the spectrum of evolutionary behavioral sciences to classify the adaptations that have been soundly supported, *as well as call attention to those which may be lacking in empirical support*. In this way, the project will help researchers debate and empirically evaluate which psychological phenomena are evolved adaptations and which are not, irrespective of formal academic boundaries (my emphasis).

A purported psychological adaptation might have evidence falling in the psychological category (as is likely to be the case for most of these purported adaptations), and evidence falling in the cross-cultural category, but lacking hunter–gatherer and physiological evidence, as well as only having moderate evidentiary depth in some categories. Further studies might be required within the existing categories of evidence or fresh investigations in the other categories of evidence, often requiring the tools and knowledge of adjacent practitioners.

Second, codifying and evaluating evolutionary psychology research should better motivate evolutionary psychologists into pursuing cooperation with researchers across the behavioural sciences spectrum. For example, reading the PsychTable might reveal that developmental evidence for a purported adaption is entirely absent. An evolutionary psychologist who champions the purported adaption in question might then actively seek out cooperation with those who specialise in child development research, to investigate rigorously whether the psychological trait develops easily, quickly and reliably.

Third, the PsychTable might better organise research efforts. The evidential standing of a purported psychological adaptation might not be as extensive or as rigorous as generally believed. Indeed, there might be strong evidence against a purported psychological adaptation. Knowing the state of play will better determine where precious research effort is best spent, which purported psychological adaptations are the most promising and which are the least promising. Indeed, 'one of the consequences of aggregating published evidence to support or refute purported EPAs [evolved psychological adaptations] will inevitably be the revelation that some or many mental mechanisms proposed and even accepted in the evolutionary psychology literature will not pass empirical muster' (*ibid.*: 314).

Fourth, the PsychTable should enable the layman to better understand the evolutionary psychology programme. Buller's (2005) underlying motivation was to correct what he took to be the layman's misunderstanding of the evidential strength of evolutionary psychology. As Buller (2005: 16) says,

> Evolutionary Psychologists have been very successful in conveying their ideas to a broader public, and I strive in this book to covey to the same broad public the other side of the story. Everyone should be able to understand the problems with Evolutionary Psychology and to understand why we must move beyond Evolutionary Psychology in order to one day achieve a better evolutionary psychology.

With the PsychTable, we don't have to rely on contentious representations—either from overly enthusiastic proponents or from overly ungenerous sceptics. Rather than being held hostage to a particular individual's experience, understanding, and representation of the literature, we can freely and easily access a much firmer and objective evaluation of the state of play, both for each purported psychological adaptation and for evolutionary psychology as a whole.

I hope the reader will feel a quiet sense of excitement about this. With a working version expected to be available in the near future, this, I believe, could be a game changer. No longer does the literature have to proceed in the all-too-predictable, Punch and Judy dialectics of 'this is a psychological adaptation', then 'but you don't have evidence from this category', then 'well we do, see such and such study in 2002, why didn't you cite that', then 'well that evidence is weak', then 'actually it's a lot better than that', then 'well what about this...', and so on. With the ever escalating amount of work done in evolutionary psychology, an accurate characterisation of its actual output is needed now more than ever. By classifying and evaluating evidence, and making this easily available in a central location, clear pathways for new research will become visible from the restless sea of research. This could significantly help foster the interdisciplinary cooperation needed to vigorously vindicate psychological adaptation claims. It is certainly something to hope for. And it may just happen.

Yes, the evidence required to vindicate completely adaptation claims is difficult to accumulate, very difficult. Nevertheless, this doesn't mean it's unlikely the appropriate evidence will ever be forthcoming. Nor should it be a reason to be uniquely sceptical of evolutionary psychology furnishing more complete evolutionary (ultimate) explanations—the issue of paucity of historical details applies whenever we seek to model evolutionary processes, and it applies as equally to non-selectionist evolutionary models as it does to selectionist models (if one claims that drift played an important role in the evolution of some inherited trait, to establish a complete explanation, one still requires the evidential particulars of ancestral population size, and so on). Furthermore, this is no different to the rest of science. In whatever branch of science that attempts to model and uncover unobservables, reality has to be fought for and won. History bears witness to many dismissing embryonic scientific developments as speculative, only to find themselves on the wrong side of history. In an entirely different context, Callender (2012) observes,

> I think it's a fool's errand to try to prejudge inquiries on the basis of being too speculative or not. History is littered with embarrassing judgments stating that such-and-such is too speculative—e.g., Newtonian gravity, quantum non-locality, relative simultaneity—only to see those ideas later vindicated. I want no part of that tradition.

The point carries over perfectly into the context of our present discussion.

The frequent strides made by the evolutionary and behavioural sciences, by geneticists, evolutionary anthropologists, archaeologists and others should give those given over to counsels of despair some pause for thought. It seems hardly a month goes by now, perhaps even hardly a week goes by, when another facet of the past is illuminated still further.

One fast-moving area of research is the reconstruction of ancestral population migration patterns and gene flows between ancestral populations, one of the five evidence categories cited by Richardson. William Blake wrote of seeing a world in a grain of sand. In a way, we can now do likewise: we can now see a world in a strand of hair. Rasmussen *et al.* (2011) sequenced the Aboriginal Australian genome from an Aboriginal Australian's strand of hair. Comparing the sequenced Aboriginal Australian genome with the DNA from other populations around the world revealed that the ancestors of modern Aboriginal Australians were the first to separate from other humans in Africa, dispersing into eastern Asia between 62,000 and 75,000 years ago, a dispersal separate from the one that gave rise to modern Europeans and Asians between 25,000 and 38,000 years ago. And on 30 August 2012, a paper in *Science* announced the successful sequence of the Denisovan genome from a tiny finger bone fragment, enabling, according to the authors of the research paper, 'detailed measurements of Denisovan and Neandertal admixture into present-day human populations, and the generation of a near-complete catalog of genetic changes that swept to high frequency in modern humans since their divergence from Denisovans' (Meyer *et al.*, 2012).

Recall from Chapter 2 that Gould (2000) asked the legitimate question of how is it possible to know in detail what our ancestors did millions of years ago, how is it possible to obtain the key information required to vindicate adapataionist explanations? Recall that after listing some of the key information needed to vindicate adaptationist explanations, he asked, 'We do not even know the original environment of our ancestors—did ancestral humans stay in one region or move about?' (2000: 120). Even as recently as a decade ago it was easy for such a question to yield rapidly to a counsel of despair. Now such a question directs our attention to and highlights the incredible advances those disciplines dedicated to reconstructing the past are making. Today even migration patterns can be inferred and reconstructed from the tiniest of fragments. And if that is so for migration, what about for the other key information categories needed to complete adaptationist explanations?

Finger fluting research is another fascinating window into the past. Finger fluting is a type of cave art, where marks have been made into clave clay with fingers. Sharpe and Van Gelder (2005, 2006) developed a methodology allowing different individuals to be distinguished on the basis of the flute measurements, even allowing the age, and sometimes even the sex, of the individuals to be determined. Intriguingly, deep inside the Rouffignac Cave, France, such art is found not only on cave walls, but also high up on cave ceilings, a few feet higher than can be done by an adult tiptoeing. Analysis of the flutes on the ceiling walls, likely to be at least 13,000 years old, reveals they were done by children, presumably lifted up by an adult or supported on an adult's shoulders.

A third instructive example: Berna *et al.* (2012) conducted microstratigraphic investigations on burned bone and ashed plant remains in the Wonderwerk Cave, South Africa. The molecular analytical technique uncovered 'unambiguous evidence' that the burning of the bones and ash took place in the cave during the early Acheulean occupation in the Lower Pleistocene, approximately a million years ago. This means ancestral populations were using fire 300,000 years earlier than previously thought. The researchers concluded, 'We believe microstratigraphic investigations at Wonderwerk cave and other early hominin sites in Asia and South and East Africa will have a significant impact in providing fundamental evidence for the appearance of use of fire and its role in hominin adaptation and evolution' (Berna *et al.*, 2012: 1220).

Furthermore, two recent research papers shed further light on the Neanderthals' behavioural activities. First, in a research paper titled 'Neandertal humeri may reflect adaptation to scraping tasks, but not spear thrusting', Shaw *et al.* (2012) examined Neanderthal arm asymmetry. As evidenced by their skeletons, Neanderthals had drastically overdeveloped upper right arms: their upper arm bones are up to 50% stronger on the right side than on the left. This arm asymmetry is significantly more pronounced than what obtains in modern humans—our upper arm bone strength asymmetry is only around 10%.

The arm asymmetry is indicative that Neanderthals were doing something laborious and repetitive, and that this represented a considerable role in their lives. Evolutionary anthropologists usually explained the pronounced arm asymmetry as the result of the Neanderthals, who were right-handed, using spears when hunting. Hunting clearly takes up a great deal of time and would have clearly been a critical activity.

Shaw *et al.* (2012: 2) proposed that other possibilities are viable and that 'testing would help to assess whether this, or any other functional hypothesis, is the most appropriate explanation for the unique morphological characteristics of Neandertals'. Side-scrapers, a tool use to scrape tissue from the underside of animal hides, are often found with Neanderthal skeletal remains. Scraping animal hides would have been a laborious and repetitive task. Accordingly, the researchers investigated whether Neanderthals' arm asymmetry is better explained as an adaptation for scraping, rather than spear-thrusting.

The researchers used electromyography, a technique for measuring the electrical activity produced by skeletal muscles, to measure muscle activity in the three deltoid (shoulder) muscles during three underhanded spear-thrusting tasks and four scraping tasks. They found that the muscle activity required to perform various spearing tasks does not explain the overdeveloped upper right arm strength, whereas the results were consistent with the muscle demands imposed by scraping activities.

While acknowledging the study is not conclusive, and that further research is needed, the researchers nevertheless stressed 'these results yield important insight into the Neandertal behavioural repertoire that aided survival throughout Pleistocene Eurasia' (Shaw *et al.*, 2012: 4). If it is indeed the case that Neanderthal arm asymmetry is better explained as an adaption to scraping tasks, or at least is a significant behaviour determining the asymmetry, then this indicates that Neanderthals spent a sizeable amount of time on daily subsistence tasks such as preparing skins, possibly for producing clothes (especially likely, given the cold conditions Neanderthals lived in). We therefore gain a richer understanding of the Neanderthals' daily activities.

Second, Hardy *et al.* (2012) extracted material entrapped in the dental plaque of five Neanderthal skeletons in the Sidrón Cave, Spain. Molecular evidence of the material revealed that, in addition to meats, the Neanderthals at the site also ate a range of cooked plants, including bitter-tasting medicinal plants. This 'suggests that the Neanderthal occupants of El Sidrón had a sophisticated knowledge of their natural surroundings which included the ability to select and use certain plants' (2012: 617).

It's astonishing, truly a marvel of recent years, that technological advances are enabling us to recover migration patterns and gene flow from a strand of hair or fragment of bone, profile individual finger fluters, establish the advent of fire-making with every greater accuracy, and recover ever more of Neanderthals' daily activities and dietary patterns. Mithen (2007: 60) discusses a different set of considerations, but the conclusion he draws is equally applicable here:

> In a very real sense, this evidence does allow us to 'directly observe what happened during human evolutionary history'. Not 'observe' in the behavioral sense of witnessing specific actions taking place, but in terms of the material products of that behavior, whether it is debris from manufacture of stone artefacts or paintings on cave walls.

It's likely that that our knowledge of the past, our knowledge of ancestral populations—population migrations, gene flows between populations and population structures—will continue to grow and mature. Multidisciplinary cooperation is an important driver of increasing such knowledge. When reconstructing our evolutionary past and the evolution of our psychology, evolutionary anthropologists, evolutionary psychologists, cognitive scientists and others should cooperate with one another. Indeed, some of the most exciting and important progress in the evolutionary sciences has arisen from the cooperation, convergence and synthesis of fields. After all, the modern evolutionary synthesis, the consensus framework of modern evolutionary biology, is the synthesis of genetics with natural selection, a dialogue between mathematical geneticists and naturalists, between those studying hereditary and those studying populations of organisms. In more recent times, molecular evolution and phylogenetics have converged to form molecular phylogenetics, which seeks to establish the rates and patterns of DNA change and to reconstruct the evolutionary trees of genes and organisms, generating fresh insights on a variety of issues, generating new insights into the past.

So we're creating ever better models of the past, illuminating the past like never before. Think of where we've come from, from the time before Darwin, when prehistory was a mystery, how much progress has been made since *On The Origin of Species* was first published a little over 150 years ago, and how much more awaits to be recovered as more and more genomes become sequenced, as computational power grows, as ever more complex demographic scenarios can be computed and then matched against sequenced genomes and other data sets, as new techniques and tools are invented allowing innovative inferences, as cross-disciplinary activities are further utilised, as evolutionary anthropologists and cognitive psychologists work together more, as the evolutionary and behavioural sciences slowly, quietly converge at various points—and as evolutionary psychology's own hypotheses become ever more sophisticated, in tandem with empirical discoveries in cognate and adjacent fields, and as its own output grows still further and stronger.

Thus, we shouldn't be ignorant of, or dismiss as hopeless, the ever-increasing strides in modelling and uncovering the past made by archaeologists, geneticists, evolutionary anthropologists and others. Such efforts give us real grounds to be cautiously optimistic about the success of gathering sufficient evidence in the long run to better complete evolutionary (ultimate) explanations—whether selectionist or non-selectionist. At the very least, such remarkable efforts offer little in the way of support for poorly thought out counsels of despair.

4.4 Conclusion

The thought, motive and practice of evolutionary psychology is that ancestral populations faced adaptive problems, that some or many of these problems were solved by specialised psychological adaptations, and that by focusing on adaptive problems and solutions we can make new discoveries in psychology. This is a programme primarily of discovery, and represents one of many possible programmes in the evolutionary behavioural sciences.

However, evolutionary psychologists at the summit of the programme, those whose views are widely disseminated and known to people outside the programme, frequently adopt and advocate tenets well beyond what's minimally needed. In itself this need not be a problem. The trouble arises when science is mixed with philosophy of nature without clear distinction, when the ideas and methods for investigating possible psychological adaptations are mixed with more ambitious, contestable tenets, when the research programme is presented with pseudo-paradigmatic or full paradigmatic trappings.

This carries substantial risks. The more unnecessary posturing, the more unnecessary controversy, the more the heart of the programme is obscured. If the root and branches are not clearly distinguished, then a sceptic might be liable to dismiss the programme root and branch, rather than just pruning back certain branches.

Framing evolutionary psychology as a heuristic project, rather than as a paradigm, therefore, has much to offer. The immediate impact is to focus more on daily practice, rather than be drawn away from daily practice. Standard objections evaporate. For example, by prizing away tenets such as a commitment to massive modularity we prize away all the controversy that tenet generates. Ditto the other tenets. Potential misunderstandings greatly reduce. Importantly, this reframing also disarms many of the criticisms based on the incompleteness of its explanations—criticism that is the source of much unease about the programme. Failure to establish all evidential particulars, either at the present moment or in the final verdict, of an evolutionary psychology explanation does not in any way undermine the heuristic value of adaptationist hypothesising.

Furthermore, with this reframing, evolutionary psychology becomes open to a wider audience. And it becomes recognised for what it is—a member of a wider constituency. This recognition is crucial if we are to have any hope of better completing adaptation explanations. Complete adaptation explanations are demanding. The sceptics are quite right in drawing attention to this. But highlighting this issue will not win for them the conclusion they seek. Crucially, we can pursue adaptationist hypothesising without having to first establish many of the evidential particulars required of a complete adaptationist explanation. Indeed, we engage in adaptationist analysis precisely as a first step in a journey, a long journey, towards establishing such details.

Evolutionary psychology explanation stands continuously exposed to demands for further evidence and details and this shapes directions of research. So yes, while complete portfolios of evidence do not yet exist, the bigger point is that evolutionary psychology can help stimulate other research programmes and disciplines into helping find such details. The forthcoming launch of the PsychTable, a classification system that seeks to codify purported psychological adaptations in one location and evaluate their evidential breadth and depth, should help facilitate this.

Thus, evolutionary psychology's role in the evolutionary and behavioural sciences is that of discovery—it's for the evolutionary and behavioural sciences collectively to recover the spectrum of evidential details. Evolutionary psychology can directly lead to new discoveries, as Richardson reluctantly acknowledges, and it has done so, as I have indicated in previous chapters, and it can potentially indirectly lead to new discoveries via stimulating other research programmes, as I have argued in this chapter. So we can, and must, shift the burden of the argument to the other side. Why not carry out evolutionary psychology as a heuristic project? The heuristic reasons I have been identifying and vindicating throughout these two chapters are, I believe, irresistible at this point. Nothing the sceptics have raised destabilises or discredits this fundamental point.

5
Restructuring the Debate

5.0 Introduction

In the previous chapter we discovered how evolutionary psychology, as a heuristically motivated project, has the potential to stimulate multi-disciplinary research activity. Multi-disciplinary research activity is born not just from a shared platform of knowledge, but also from an attitude of respect, a conviction that other disciplines have something important to say. For many years now these two factors have been strikingly absent in the debate. One of the stumbling blocks to the possibility of multi-disciplinary research on purported pyschological adaptations being realised—the lack of a central database of purported psychological adaptations—will hopefully be removed sometime in the near future. But the other stumbling block, the polarisation of the debate on evolutionary psychology, remains as entrenched and powerful today as it ever has been. In light of the previous chapters, I believe we now have the resources to make sense of the nature and character of the heated debate, that we can not only honour the motivations of both sides of the debate, but also highlight where both have gone wrong, to highlight their strategic and argumentative mistakes, and to thereby depolarise and reconfigure the debate, enabling the debate to finally move on and for genuine multi-disciplinary research to gain momentum. The fruits of doing so are too great to spoil by continuing the status quo.

How can such an entrenched and heated debate depolarise? To depolarise the debate is to champion two fundamental strategic shifts: it requires arguing that sceptics should recognise the heuristics of the programme and to consent to the legitimacy and reasonableness of the heuristic enterprise, and it requires arguing that prominent evolutionary psychologists should not mask those heuristics with theoretical excess.

The lesson drawn from the previous chapters should be now clear: one can contest the paradigmatic trappings of evolutionary psychology and one can, and indeed should, question the evidential status of evolutionary psychology explanations and yet one can also recognise and consent to the heuristic goal of the programme. The sceptics have focused most of their endeavours on paradigm and explanatory or evidence issues and comparatively little on the heuristic dimension. In 5.1, I call this their 'fundamental interpretative mistake'. In 5.2, I argue that prominent evolutionary psychologists have largely encouraged this fundamental mistake: evolutionary psychology's purported tenets and even its very name have created a powerful expectation about the programme that bears little relation to the daily activity of the programme. It should come as no surprise, therefore, that sceptics have emphasised features of evolutionary psychology that have little bearing on actual practice.

As evolutionary psychology is in the business of generating and testing hypotheses about psychological mechanisms, then it is not only eminently reasonable, but it should also be welcomed by all. If both sides endorsed this picture, this reframed or streamlined evolutionary psychology, much of the heat of the debate can dissipate and the debate itself can move onto new issues and opportunities, rather than endlessly rehearsing and updating well-entrenched arguments.

Even after this, there remains one final challenge to engage with. If we are to establish a reasonable position on evolutionary psychology, one that all reasonable parties can consent to, we need to also be aware of, articulate and challenge a relatively new, more moderate scepticism, but one which can be just as subversive as the hard scepticism. This position recognises that evolutionary psychology has made genuine contributions but that in a sense its best days are over, and that its time for its practitioners to move on, to 'modernise'. I motivate and challenge this position in 5.3 and 5.4, where I argue that calls to modernise evolutionary psychology are based on a failure to recognise or appreciate its unique role in the behavioural sciences. I conclude by advocating that strong sceptics, prominent champions of the programme, and the milder sceptics, respectively, focus on, refocus on and revalue the heuristics of the programme.

5.1 Focusing on Explanation: The Fundamental Interpretative Mistake

The sceptics' fundamental interpretative mistake, the mistake that cascades a host of other mistakes, is to overlook evolutionary psychology's heuristics, to instead focus too strongly on the evidential incompleteness of its explanations. In other words, to get the wrong end of the stick. Evolutionary psychology's scientific credibility does not derive from its ability to accommodate known social phenomena within an adaptationist framework, in its ability to advance plausible reasons for behaviour we already know about, but in its ability—its rich ability—to generate novel predictions about behaviours yet to be investigated. Here we are to find evolutionary psychology's methodological vindication. Here we are to find evolutionary psychology's value and function in the evolutionary behavioural sciences.

In light of our discussion in the previous chapters, it might now seem a little striking that Richardson (2007) writes in considerable detail for over 200 pages on evolutionary psychology without seriously recognising or engaging with the possibility that evolutionary psychology primarily functions heuristically. Relying on his book alone, one might perhaps not realise that evolutionary psychology has generated batteries of novel predictions, surprising predictions and that many have been successfully corroborated. For example, when discussing Buss's research on jealousy, Richardson (2007: 88) observes that Buss only makes general appeals to Pleistocene conditions; particular historical conditions of human evolution are absent. He furthermore notes that the only evidence Buss offers is evidence from psychological experiments. Richardson is absolutely right about that: the evolutionary history cited is thin, certainly not the rich details Richardson rightly expects of a complete adaptationist explanation, and the psychological evidence cited can be accounted for by non-adaptationist explanations. But what Richardson doesn't mention is that Buss didn't merely appeal to existing psychological evidence but rather he created fresh psychological evidence. He expanded the explanandum of jealousy.

Buller (2005) presents a focused and forensic challenge against evolutionary psychology. Spanning over 500 pages, *Adapting Minds* is a sustained and influential compliant and case against the programme, codifying, it seems, every significant criticism ever raised against it. And yet in all those hundreds of pages, there is little recognition or engagement with evolutionary psychology's heuristics. Kitcher and others call evolutionary psychology 'sociobiology reborn'. 'Evolutionary psychology turns out to be pop sociobiology with a fig leaf' (Kitcher and Vickers, 2003: 334). But to think of evolutionary psychology as sociobiology reborn is, I believe, to betray a profound misunderstanding of the programme, to signal that the basic picture of the programme hasn't been grasped. It is to miss that evolutionary psychology is using ultimate considerations to shed light on proximate design, to instead take it as just another attempt to give evolutionary stories for extant behaviours. And finally, as we saw in Chapter 2, Schulz (2011) is one of the few sceptics who recognize the possibility of evolutionary psychology being heuristic. But despite seeking to examine whether evolutionary psychology is heuristic in practice, the heuristics of the research programme were not explicated in any detail. Had that been done I expect it would have been significantly easier to see just how frequently evolutionary psychology can and does operate heuristically in practice, just as we have in this book.

Evolutionary psychology's heuristics should occupy centre stage in any evaluation of evolutionary psychology. But such considerations are absent or overlooked in the sceptical literature. That many sceptics fail to see the basic picture of what evolutionary psychology does in practice is troubling. While the strident tones and extreme dismissal of evolutionary psychology found in these works might make for entertaining narratives, and they truly do, they don't make for a meaningful portrayal of the experience of hundreds of practitioners of the programme. On the contrary, this basic mistake fuels a seriously distorted picture of the programme—in works bearing unfortunate titles like 'Evolutionary Psychology as Maladapted Psychology'.

A sceptical literature based on a basic interpretative mistake can only generate, and sustain itself by, a dizzying succession of other mistakes. I identify six such mistakes. First, by overlooking its heuristics, sceptics inevitably develop a cartoon understanding of evolutionary psychology's methodology. As we've seen, evolutionary psychology's methodological framework is vibrant. Hypotheses can be continually tested, refined, modified or abandoned. Far from being simple, it's a rich framework for generating novel hypotheses. Yet by overlooking the heuristics, sceptics fail to recognise, let alone engage, with this sophisticated methodology. Take Kitcher and Vickers (2003) as a dramatic example. Kitcher and Vickers call evolutionary psychology's history 'dismal' and then proceed to describe its methodology. This is something worth quoting in full because the quote captures in an illuminating nutshell the focus and motivation of their intense critique:

> Here's a recipe for winning fame and fortune as an architect of the new-and-improved human sciences. First, make a bundle of claims to the effect that certain features are universal among human beings, or

> among human males, or among human females. Next, couple each claim with a story of how the pertinent features were advantageous for primitive hominids, or males, or females, as they faced whatever challenges you take to have been prevalent in some lightly sketched savannah environment. (Don't worry that your knowledge of past environments is rather thin—Be creative!) Finally, account that each feature in the bundle has been shaped by natural selection, and so corresponds to something very deep in human nature (male human nature, female human nature), something that may be overlain with a veneer of culture but that molds our behavior and the forms of our societies. Accompany everything with hymns to the genius of Darwin, broadsides against 'blank state' views of the human mind, and vigorous denunciations of the lack of rigor and clarity that has hitherto reigned in the human sciences (Kitcher and Vickers, 2003: 333).

As you can see, this captures the picture of evolutionary psychology irritating hard sceptics, motivating his or her efforts to expose it for what it is perceived to be. And you can now see the unhealthy distance between this depiction of the methodology and the depiction of the methodology as advanced in this book. See how the methodology is conceived as obviously and fatally flawed, as crude as it is potentially dangerous. Sceptics have a crude picture of evolutionary psychology methodology. It is therefore unsurprising that they find evolutionary psychology crude.

Second, overlooking the heuristics leads to a stark assessment of the programme, sparing in praise, lavish in condemnation. Evolutionary psychology has had a profound influence on charting efforts in psychology. For example, Cosmides and Tooby's work on the Wason selection task ignited much research into a previously ignored and overlooked area. Alternative explanations were proposed. Alternative explanations were experimentally tested. New experimental knowledge was won. Even if such research stands incomplete, and even if their interpretation of the results, their explanation, turns out to be false, evolutionary psychology has been instrumental in increasing knowledge in this area, as it has done so in multiple other areas - and it is likely to still do yet further. Evolutionary psychologists spend large amounts of time engaging in selectionist and engineering reasoning, formulating hypotheses, carefully teasing out surprising novel predictions, spending time constructing psychological tests, recruiting participants, collecting data, evaluating data, submitting their work to peer review, indicating where further lines of research are needed, further tests that need to be made—for which the sceptics denounce them for peddling in just-so stories, and greet their hard-won gains with the unforgettable line, 'Some may eventually yield to empirical discoveries. If this is so, then so be it' (Richardson 2007: 38).

Third, overlooking the heuristics leads sceptics to not recognising how evolutionary psychology merely follows normal scientific practice of modelling unobservables. Of course, behaviours don't fossilise. We don't dig around a cave, pick up a rock and say 'ah, here's jealousy'. We'll never unearth a coalition fossil or a cheater detection fossil. But that's not a licence for pessimism about recovering past selection pressures. Often the pessimism about recovering the past boils down to little more than a complaint that we don't have a time machine, a point also noted by Campbell (2006: 90), who says 'some have argued that our inability to time travel to the environment of evolutionary adaptedness (EEA) means that we can never "prove" which environmental problem the adaptation was designed to solve'.

The point is that we can create models or hypotheses of the unobservable past, hypotheses that produce novel predictions, surprising predictions about observables—and this is merely the normal practice, the bread-and-butter, of science. We can't see many unobservables. But we don't give up on trying to understand what obtains beyond observation. We don't fall into counsels of despair about knowing the origins of the universe or what exists at the subatomic levels. By carefully building possible scenarios, by seeing what fits, we can eventually build credible models. Incidentally, on this criterion, evolutionary psychology is actually in a better position than M-theory, which is commonly noted for not generating novel predictions.

Fourth, because they are largely blind to its ability to generate novel predictions, sceptics simply don't see on what basis one could bet on or tentatively accept evolutionary psychology explanations. Even if there's currently only an incomplete portfolio of historical evidence, and even if ultimately that portfolio will remain incomplete, there are still grounds for accepting individual evolutionary psychology explanations over rival explanations. Despite lacking a range of evidential details, we're entitled to favour or accept evolutionary psychology explanations if they outcompete rival explanations by explaining the same range of facts as rival explanations while simultaneously making surprising predictions that the other explanations are blind to.

As a general mode of scientific procedure, we are inclined to accept hypotheses and explanations insofar as they make successful predictions of new data. Lipton (1991) observed that when Dmitri Mendeleyev initially formulated his periodic table of elements, which accommodated the 60 elements known at the time, this was only mildly impressive. But when this reorganisation led to the further novel prediction, subsequently corroborated, of two further elements, this was recognised and rewarded with much acclaim. 'Sixty accommodations paled next to two predictions' (1991: 135).

We know that explanations are subject to underdetermination—many rival explanations can accommodate a given phenomenon. And in the social sciences, this is not just an abstract possibility, but an actuality: accommodations in the behavioural and social sciences are ten a penny. Therefore, in the behavioural and social sciences, predictive success is usually a more compelling basis for evaluating rival explanations. If an evolutionary psychology hypothesis has made striking predictions, surprising predictions, highly counter-intuitive ones, which subsequently enjoy a collision of evidence from a variety of sources, then we're warranted to prefer or accept such an explanation, even if we lack all the evidential particulars. Nothing special here.

Imagine if someone came along and stipulated what the 'complete evidence' for the big bang theory would be. Suppose further that this sceptic claimed we don't satisfy every evidential particular, that we're unlikely to do so, and then concluded from this that the big bang theory is unbridled speculation. We would find that incredulous. We would point out that we can conceive or evoke a range of hypotheses, each of which make alternative predictions about observables, over various measurements and indicators. One of them, the big bang theory, offers better predictions and the best fit with data. On that basis, we're entitled to favour or accept the big bang explanation over rival explanations. Yes, alternative explanations are logically possible, and further research is always possible – but this is no different to the rest of science. Ultimately, we wouldn't take the sceptic seriously. We would consider her to be adopting a scientifically irresponsible attitude. Yet in the highly polarised arena of evolutionary psychology discussion, arguments parallel to this are put forward that would be laughed out in other contexts.

Of course, evolutionary psychology-led discoveries can still be accommodated by non-selectionist explanations. But this is neither peculiar nor interesting: all predictively successful hypotheses can be contested, via disputing the auxiliary assumptions that mediate between hypothesis and prediction such as the validity and execution of measurement apparatus or probing and leveraging the possibility of confounding factors.

Furthermore, one can usually revise an explanation to accommodate a new fact. So there's usually always a generous amount of wiggle room to save a rival explanation. And here a distinction becomes crucial: the important point is to distinguish between programmes that lead to empirical discovery, and programmes that simply accommodate the fresh discoveries of other programmes. For any given explanandum where rival explanations do battle, ask which research programme is leading the way, expanding explanadums, and which is merely playing catch-up.

The second half of Buller (2005) looks at several case studies of evolutionary psychology, pointing out possible alternative explanations throughout or calling into question the interpretation of results, without sufficiently highlighting that all explanations and results are subject to underdetermination—indeed, I don't recall any mention of the problem of underdetermination, despite underdetermination being heavily leveraged throughout (and which expose his argument to similar counter manoeuvres—see Cosmides *et al.* (2005), Buss and Haselton (2005), Daly and Wilson (2005b), and Delton *et al.* (2006), who dispute various evidential details and interpretations utilised by Buller).

And, of course, some, many or all of evolutionary psychology explanations might turn out wrong and eventually be discarded. But again, how is that any different from the rest of science? This possibility is not unique to evolutionary psychology—it's ubiquitous to all scientific endeavour. One can never have a guarantee that one's research programme will work, and one never has a full explanation at the beginning. It took Newtonians decades if not centuries to provide a more or less complete explanation of the solar system's behaviour, and eventually this explanation was displaced (the famous case of Mercury's perihelion). That is simply how science progresses.

Evolutionary psychologists themselves are alert to this. Recall from Chapter 3 that Sell, Tooby and Cosmides postulate a neurocognitive programme that automatically calculates welfare tradeoff ratios (WTRs). Sell *et al.* claim that 'WTRs are not just post hoc theoretical constructs, but exist as real neurocognitive elements within the human motivational architecture, playing a role in decision-making' (2009: 15074). Have Sell *et al.* conclusively established this? Barkow nurses a degree of caution: 'No, not unless one has a great deal of faith in construct validity: accurate prediction can make a construct useful but is not, in my opinion, sufficient for one to be as certain of its ontological status as are Sell, Tooby, and Cosmides' (Barkow, 2009: 14743–4). Barkow thinks that we shouldn't be too entangled with the question of whether the brain 'really' is a biological computer. The more productive question for Barkow is whether the biocomputational approach leads to theories and research that increases our understanding of human behaviour. Barkow believes that it is, at least judging by 'the myriad' of research evolutionary psychology produces. But 'at this point in history', the WTR, as well as the biocomputation perspective that underlies it, is only a useful construct, a productive way of cashing out an evolutionary perspective. 'Eventually', Barkow continues, 'the Cosmides and Tooby biocomputational approach may be replaced by more complex theoretical formulations, but in the meantime, as the Bohr model of the hydrogen atom did in its day for particle physics, that approach is moving us forward and does not appear to have any demonstrably superior competition' (*ibid.*: 14744).

Fifth, focusing too strongly on evidential issues, and too little on the heuristics, means it's only a short journey from claiming evolutionary psychology research is merely speculation to saying its dangerous speculation. Evolutionary psychology hypotheses are routinely condemned as being just so stories of the worst sort: of recasting stereotypes into an evolutionary mould. Evolutionary psychologists are suspected of infusing various biases—sexist biases, ideological biases—into their work, building into their hypotheses of their own views of society. 'Evolutionary accounts of human behavior often involve constructing the past from the present. For this reason, such accounts are highly susceptible to ideological biases and cultural blinders' (Caporael and Brewer, 1990: 287). And as Dupré puts it, 'Reading these accounts of male–female relations, one is struck by a mixture of the stereotypic, the outrageous, and the banal' (2001: 54).

Unsurprisingly, therefore, the sceptical tradition sometimes slides into bad language and naked anger. Daly and Wilson note that sceptics' incivility is a 'disturbing and sometimes perplexing element... They are not just sceptical, they are angry, and we are still not entirely sure what they are angry about' (2007: 396). As we've seen, Kitcher and Vickers (2003) call evolutionary psychologists 'pop sociobiologists' and 'reformed pop sociobiologists'. Gray *et al.* (2003) speak of their 'attack' on evolutionary psychology. Using disparaging terms is not by any means the worse thing one do, but it does illustrate how emotions have adulterated the sceptical literature.

I'm not claiming, and need not claim, that it's impossible for evolutionary psychologists to work various ideological biases into their work, but the assertion seems way overblown. Beyond possibly a few examples, there's little evidence of this, and some evidence they don't. For example, Tybur *et al.* (2007) surveyed the political attitudes of 168 psychology PhD students in the USA, 137 of whom were self-identified as non-adaptationists and 31 of whom were self-identified as adaptationists. The survey revealed that adaptationist PhD students in psychology are no more politically conservative than non-adaptationist graduate students, and are less politically conservative than US citizens in general. Furthermore, and more importantly, ideas take on lives of their own, beyond the motivations of their authors. Even if an evolutionary psychologist projected suspect stereotypes and ideology into his or her work, such hypotheses are unlikely to survive testing, scrutiny and the peer-review process.

Sixth, by getting the wrong end of the stick, sceptics of evolutionary psychology become isolated and adrift from developments in evolutionary psychology. The PsychTable we looked at in the previous chapter shows how fast moving evolutionary psychology and the evolutionary behavioural sciences are, and how increasingly disconnected and outdated the sceptical literature is. Indeed, if someone had the time to perform one of those word clouds on typical specimens of the sceptical literature, I'd venture certain word- and phrase-clusters would stand out above all others: 'adaptations aren't optimal', 'not all traits are adaptations', 'we don't know the past', 'speculation', 'just so story', 'cheater detection', 'rape adaptation', 'where are the genes', 'what about this possibility', 'what about that possibility', 'other explanations possible'. Especially 'other explanations possible'—as if evolutionary psychology is uniquely subject to underdetermination. This unhealthy focus on just a few cases and the running of underdetermination arguments makes the sceptical literature sound a little like it's stuck on autoplay.

Despite sceptics' efforts, the ever escalating proliferation of evolutionary psychology work strongly suggests the programme is rooted and growing, slowly branching out to flower fruitfully in sometimes surprising twists and turns. Evolutionary psychology has identified possible psychological adaptations underpinning kin relationships, detection of cheating, language acquisition, theory of mind, folk physics, face recognition, mate preferences, cooperation and specific emotions, such as fear, jealousy and disgust. It has explored possible informational inputs of these mechanisms and their transformations into behavioural outputs. And, as we saw in Chapter 3, evolutionary psychology is not limited to these topics. We looked at an example of evolutionary psychology making a novel prediction about a non-social behaviour—the previously unknown descent illusion. We also noted that the distinction between adaptations and their by-products potentially enables us to apply adaptationist thinking to a broad range of phenomena. A striking example of a proposed by-product of a single psychological adaptation is psychology of race. Kurzban *et al.* (2001) reasoned that it is unlikely ancestral populations encountered different races—the psychology of race might well be a by-product of the psychology of coalitions. We also looked at an example of a purported by-product of multiple psychological adaptations. Religion is costly, counterfactual and counterintuitive. How, then, could it have arisen and persisted? According to the by-product approach, although religion is not an adaptation, the cognitive systems that converge to give rise to religion as a by-product are adaptations. This is promising and illuminating.

Indeed, evolutionary psychologists can cite increasing volumes of evolutionary psychology work that goes well beyond the bread-and-butter of mate choice, attractiveness and cooperation. For example, in 2012, the edited volume *Applied Evolutionary Psychology*, published by Oxford University Press, contained 25 papers on applications of evolutionary psychology on a diverse range of topics, including evolutionary psychology work on academic learning (Geary, 2012); business and management (Nicholson, 2012); marketing and communication (Griskevicius *et al.*, 2012); intergroup prejudice (Park, 2012); psychology of mass politics (Petersen, 2012); and even perfume design (Roberts and Havlicek, 2012).

So evolutionary psychology can cogently expand far beyond its initial topics of mating, parenting, kinship, status and aggression, which were early topics, obvious topics and move beyond those topics in surprising, sophisticated and useful ways—not in a crass takeover fashion, not in a way that necessarily displaces other explanations, but in a way that genuinely contributes to understanding. These cases suggest that many new and diverse applications of evolutionary psychology are possible. Consider the possibilities that are ahead. Consider how many unknown features of known mechanisms await discovery. Consider how many unknown psychological mechanisms await to be hypothesised and explored. Consider what kind of phenomena could be by-products of single adaptations—and how many more could be by-products of multiple adaptations. Evolutionary psychology as a heuristic is too promising a tool to abandon in the face of counsels of despair.

If evolutionary psychology really was about making commitments to particular conceptions of evolution, to particular conceptions of the mind, to advancing skeletal explanations as full explanations, to forcing out other explanatory possibilities, and demanding social science and public policy to reconfigure around it, then the sceptical case should be allowed to stand. Insofar as evolutionary psychology is presented as a paradigm, and figures in the minds of its opponents as such, the sceptical case is to be honoured. The sceptical case is surely right that paradigmatic evolutionary psychology overreaches itself and breaks down at various points. It is advertised as something much more than it can deliver.

But as we have seen not only does this neglect the heuristic dimension, but it also neglects the possibility that evolutionary psychology is better characterised along such a dimension, in both its theoretical and methodological foundations and in its typical practice. And so I believe it's time for hard sceptics to soften their stance, to take down their banners and walk away from their entrenchment. In the past, when the heuristic case was unknown or opaque, a wholesale dismissal of evolutionary psychology could be understandable. But in light of our explication and discussion those who dismiss evolutionary psychology wholesale can no longer entitle themselves to that dismissal.

And the hardened position, the complete dismissal of evolutionary psychology, must be abandoned, for this type of scepticism, despite its honourable motivations, despite its attempt to enforce scientific standards, can unwittingly and ironically lead to a profoundly anti-scientific attitude, albeit localised to evolutionary psychology. Because of the existing framework of the debate, sceptics seem to end up demanding that evolutionary psychology explanations almost immediately satisfy heavy evidential demands. If such a demand were respected, the research programme could not get off the ground, and we would be robbed of important insights. So ultimately this scepticism can be dangerous, threatening to discredit a legitimate and important research activity that has unfortunately been bundled as something much more than it actually is.

Of course, the sceptic is entitled to remain sceptical of the ultimate empirical prospects of evolutionary psychology hypotheses, as well as the ultimate prospects of establishing complete ultimate explanations for proximate mechanisms. Fair enough. Perhaps she's making a bad bet, and as the evidence comes in she will have to weaken her resolve. Or perhaps she's right. But that is neither here nor there—even if many of evolutionary psychology's explanations are displaced in the long-run, even if we ultimately reject many of them and accept other explanations for discoveries made by selectionist hypotheses, pursuing adaptationist questions in psychology is still a legitimate and valuable project.

Regardless, the hard sceptic can walk away from the 'complete dismissal' position with honour: she said some right things, things that needed to be articulated, in particular placing legitimate demands on what constitutes a good adaptationist explanation.

5.2 Towards a Streamlined Evolutionary Psychology

If we want sceptics to focus on the heuristics of the programme, then those heuristics shouldn't be obscured or overshadowed. As we saw earlier, prominent evolutionary psychologists frequently adopt positions far in excess of what is required. In itself this is not a problem. The problem arises when such excess tenets are presented, either by accident or by design, as part-and-parcel of the programme—or what's necessary for the programme. I mentioned this risk before, but it's clear this risk has been realised. The overlooking of the heuristics runs in close parallel with the overshadowing of the heuristics.

5.2.1 On the Very Name 'Evolutionary Psychology'

Expectations are powerful. They profoundly shape how we understand and evaluate something. Suppose we order some product and have a ridiculously high expectation about it. It arrives and despite it being excellent in many ways it doesn't quite meet expectation. Disappointment inevitably results, despite the excellence of the product. In a similar fashion, consider academic essays or papers. A key task of an academic essay or paper is to establish and manage expectations. As many students find out early on in their academic studies, overpromising what you can deliver or signalling that an essay will cover one issue but actually covers another issue will inevitably cost marks.

Names create expectations. To name something is to create an expectation about what it is or what it does. The right name can create the right expectation. Likewise, names can also create wrong expectations, inevitably leading to traps of misunderstanding. A nice example of a name creating expectations misaligned with actuality is the 'London School of Economics' (the LSE). Even when stretched out to its full incorporated title the 'London School of Economics and Political Science', it can still give a misleading impression about what its remit is, creating an expectation that the School covers only economics and political science. In fact, the School covers a range of the social sciences, such as sociology and history, even philosophy and law, and students can specialise in these disciplines without ever entering an economics class. Yet countless times I have received confused looks, even from conference attendees, when I mention my PhD in Philosophy is from the London School of Economics. Indeed, at one international seminar, a Nobel laureate *insisted* I must have expertise in economics in virtue of my LSE affiliation, despite my protestations to the contrary. It took a good few minutes to correct his (quite understandable) expectation.

Just think about that for a moment: despite not knowing anything about me, and despite my initial replies, the person in question insisted that I have or should have a certain knowledge and skill set, exclusively on the basis of the institutional affiliation printed on my badge. Such is the power of names and expectations. Of course, this is a relatively harmless example: to correct for the expectation in this case takes a few minutes or a browse through the internet. But what if an entire research programme or disciple has been misnamed—misnamed in the sense of creating the wrong expectation? Can you imagine the kind of problems that would cause? Someone hearing of the research programme will expect the programme to conform to whatever is associated with its name. Naturally, they will be confused, disappointed, believing the programme to be deficient in various ways, and ultimately mark it down, much like a teacher does with a student essay failing to meet expectations.

Do you see what's coming? You probably can. 'Evolutionary psychology' clearly creates the wrong expectation concerning its practice. Naming the programme of focusing on adaptive problems and solutions 'evolutionary psychology' was an unfortunate move because 'evolutionary psychology' can be taken to cover the full range of the evolutional behavioural sciences. As Daly and Wilson (1999: 509) note,

> For present purposes, HEP [human evolutionary psychology] encompasses work by nonpsychologists, including even those who have deliberately differentiated themselves from 'evolutionary psychology' as 'evolutionary anthropologists', 'human sociobiologists' and 'human behavioural ecologists'. These approaches are all 'evolutionary' by virtue of their adaptationist, selectionist conceptual framework, and they are all 'psychological' to the degree that they focus on how people acquire and evaluate information and how they use that information in behavioural decision making.

As the name 'evolutionary psychology' suggests the programme covers much more than it actually does, this has inevitably caused confusion. Kurzban (2010) picks up on this, noting that many critics of evolutionary psychology think that evolutionary psychology hypotheses are concerned with evolutionary history and phylogeny, not recognising or understanding the logic of adaptationism and its central place in the discipline, and he suggests this is perhaps because such critics have been misled by the presence of the word 'evolutionary' in discipline's name.

The literature has struggled with one approach out of many being labelled 'evolutionary psychology', getting itself into semantic gymnastics to capture the distinction between evolutionary psychology as the practice of focusing on adaptive problems and solutions, and evolutionary psychology as the broad field of applying evolutionary thought to behaviour. The most common distinction to make is between 'Evolutionary Psychology' and 'evolutionary psychology', a distinction generally credited to Buller (2005)—the former referring to evolutionary psychology in our sense, the latter to referring to the broad field of applying evolutionary perspectives to behavioural phenomena. However, this distinction can be slippery: authors who adopt it can easily drop the distinction midway through a paper, or slide between the distinctions, leaving the reader uncertain as to the proper target of the criticism. Furthermore, authors can be inconsistent with this distinction between publications (e.g. Laland and Brown (2011a) use 'evolutionary psychology' for the programme, while Laland and Brown (2011b) use 'Evolutionary Psychology' for the programme and 'evolutionary psychology' to mean the broader field).

Distinguishing between 'narrow evolutionary psychology' and 'broad evolutionary psychology' is another common convention—but 'narrow' has a negative overtone (who wants to be narrow minded?). Naming evolutionary psychology 'the Santa Barbara school' is not ideal either. After all, Donald Symons and Martin Daly and Margo Wilson were utilising this style of reasoning long before Cosmides and Tooby, and one doesn't need Santa Barbara pedigree to deploy such tools. The 'High Church' and 'Low Church' distinction is also sometimes used in the literature, but the obvious religious associations make this inappropriate—and especially troublesome for those raised in Anglicanism.

The semantic disaster reaches its climax in a passage in Gray *et al.* (2003). The paper, which already gives a hint of the semantic gymnastics with its title 'Evolutionary Psychology and the Challenge of Adaptive Explanation' (original underscore), informs the reader:

> Before proceeding any further we should emphasize that the target of our critique here is *not* a broad, comparative evolutionary approach

> to psychology (evolutionary psychology or 'evolutionary psychology in the round'. ... Instead, our attack is confined to the specific program of Evolutionary Psychology associated with the 'Santa Barbara church of psychology' (Gray *et al.*, 2003: 248; original underscore and emphasis).

As the term 'evolutionary psychology' has such a multitude of possible references, and potentially covers such a complexity of subjects, the main proponents of the programme should not, in my judgement, have permitted themselves the luxury of adopting it for their particular style of research. Instead of being baptised with that name, the programme of focusing on adaptive problems and solutions would have been better served with a name like 'psychological adaptationism'—or, even more accurately, 'heuristic methodological psychological adaptationism', though, admittedly, that would have been a bit of a mouthful. Perhaps just 'heuristic evolutionary psychology' would have done the job.

An ounce of prevention is worth a pound of cure. I believe not a small measure of the confusion and hostility in the literature could have been averted had the programme been more accurately labelled from the start. Consider if evolutionary psychology was launched as 'heuristic evolutionary psychology' or 'heuristic psychological adaptationism'. Would the methodological and evidential objections really have been pressed so firmly against the evolutionary psychology had it been clear from the start that such activity is only the beginning of formulating evolutionary explanations, not the endpoint? I think not. Imagine where evolutionary psychology and the evolutionary and behavioural sciences would be now.

So should evolutionary psychology be renamed? Probably not. Probably too late now. Dupré speaks of 'the sect that has succeeded in co-opting the term 'evolutionary psychology' (2003: 112). As well as being yet another example of the unfortunate language that has disfigured the debate, it's also an illustration of the entrenched association of the broad river with a particular tributary. Indeed, the term has become so strongly associated with the particular approach explicated in this book that researchers who would semantically be entitled to the term shy away from it, calling their activities something else entirely such as 'human behavioural ecology'. There are already too many alternative labels floating about, attempting but failing to get a better cognitive grip on the confusion arising from such a *prima facie* broad term being linked with such a particular approach. I shan't add to the confusion by using fresh terms. As 'evolutionary psychology' is now associated too strongly with one approach, despite the cognitive mischief that causes, I believe the best distinction to make is between 'evolutionary psychology' for the programme and the 'evolutionary behavioural sciences' (or 'human evolutionary behavioural sciences') for the field in general, a convention adopted by, among others, Sear *et al.* (2007) and Machery and Cohen (2012).

5.2.2 Distinguishing Research Agenda and Unification Agenda

We saw back in Chapter 1 that evolutionary psychology is presented as an attempt to unify psychology within a single framework—indeed, not only an attempt to unify psychology, but also an attempt to unify the behavioural sciences in their entirety. Such grand claims made by those at the summit of the programme have clearly captivated the eyes of sceptical hawks. We already saw in Chapter 1 that Buller recognises that leading evolutionary psychologists present the programme in ambitious ways, as 'a Grand Unified Theory of the structure and evolution of the human mind' (2005: 481). Richardson also notes, 'Evolutionary psychology does have as a core part of its agenda the overthrow of much of contemporary social science and psychology' (2007: 36). Gould (1997) is characteristically scornful: 'Moreover, a larger group of strict constructionists are now engaged in an almost mordantly self-conscious effort to "revolutionize" the study of human behavior along a Darwinian straight and narrow under the name of "evolutionary psychology"'. And, as we saw earlier, Kitcher and Vickers call evolutionary psychology 'dismal' and proceed to describe its methodology. I shan't quote paragraph again in its entirety, just the beginning and the end:

> Here's a recipe for winning fame and fortune as an architect of the new-and improved human sciences. ... [Four sentences caricaturing methodology] ... Accompany everything with hymns to the genius of Darwin, broadsides against 'blank state' views of the human mind, and vigorous denunciations of the lack of rigor and clarity that has hitherto reigned in the human sciences (Kitcher and Vickers, 2003: 333).

Notice how the methodology of the programme and the grand gambit for unification of the behavioural sciences are seamlessly woven together as one narrative. Kitcher and Vickers begin with the unification gambit, then talk about (their construction of) the methodology, and then return to the gambit. If leading evolutionary psychologists had adequately distinguished between research and unification agendas in the first place, instead of indiscriminately bundling them together, perhaps the sceptics would have done likewise. Leading evolutionary psychologists should either sharply distinguish the adaptationist research programme from their own personal ambitions for adaptationism unifying the social sciences—or simply dispense with the notion altogether, as it can't possibly live up to its apparent billing.

5.2.3 Distinguishing Research Agenda and Public Policy Agenda

Fairness matters. When judging a research programme, all things being equal, we shouldn't set the bar higher than we would judge other research programmes. And yet there is an unshakable feeling that evolutionary psychology work is frequently held to standards far higher than those applied to adjacent disciplines and other social science disciplines. Kurzban (2010) articulates this nicely. As an example, he points out research by Baumeister and colleagues. Baumeister *et al.* (2007) propose the strength model of self-control. Under this model, the exercise of self-control is held to depend on a limited resource, which, when depleted, reduces our capacity for further self-control. The metaphor driving this model is that of exercising muscle: exercising muscles requires energy, energy that can be deployed, thereby leading to fatigue. Kurzban isn't impressed by this research: he believes the model to be a pre-enlightenment one, a belief in a kind of energy, 'psychology's own phlogiston', which he believes is now absurd in light of the computational theory of mind. And yet it hasn't received anywhere near the degree of scrutiny and scepticism evolutionary psychology has. From this, Kurzban reaches a depressing conclusion: even if research fundamentally clashes with best evidence, with what we know, this alone will not trigger the kind of scepticism that regularly greets evolutionary functional analysis in psychology.

The point about pressing standards more harshly against evolutionary psychology than other disciplines is a reasonable point. It's not exactly unheard of for the social sciences to give rise to weak and questionable work. Even if one believes that evolutionary psychology work is weak or questionable, why single evolutionary psychology work out for special focus?

In their valedictory remarks, Kitcher and Vickers (2003) identify the objection that much work in the behavioural and social sciences is questionable, and therefore we shouldn't be especially upset by problems with evolutionary psychology work. In response, Kitcher and Vickers justify the higher degree of scepticism levelled against evolutionary psychology work by appealing to efforts by evolutionary psychologists to draw out public policy implications of their work:

> It's not incumbent on scientific researchers to offer policy suggestions, but some recent pop sociobiologists—including Thornhill and Palmer—have defended their proposals about human nature by declaring that they can help resolve urgent social issues. Even though we conceded that they have good intentions, that they want to help decease the incidence of rape, it's hard to avoid the judgement that Thornhill and Palmer's suggestions, where not banal, will do little good. Given the speculative character of their Darwinizing and the elusiveness of their proposals, even their inability to recognize crucial issues, policies influenced by their text might well make matters worse (2003: 351).

Recommending public policy interventions on the basis of incomplete explanations seems to particularly irritate a number of sceptics. Hence, the impression of evolutionary psychology as a political agenda, one to be challenged: 'Evolutionary psychology is not only a new science, it is a vision of morality and social order, a guide to moral behavior and policy agendas' (Nelkin, 2000: 20). As Thornhill and Palmer sought to draw out policy implications for their work, this perhaps accounts for why the rape adaptation hypothesis has attracted a degree of attention in the sceptical literature far out of proportion than its presence in the evolutionary psychology literature warrants.

Of course, it's clear that evolutionary psychology work might eventually provide insights relevant to public policy. First, on the strength of evolutionary psychology work, given how a particular psychological adaptation or a set of psychological adaptations are understood to operate in our current environment, one could identify existing public policies and interventions that are unlikely to be effective. Second, the reverse side of the coin: on the strength of evolutionary psychology work, given how a particular psychological adaptation or a set of psychological adaptations are understood to operate in our current environment, one could identify—and engineer—policies and interventions likely to be successful. As psychological adaptations generate a variety of behaviours, depending on what environmental cues are being received, one could propose environmental setups that provide the right inputs, the right developmental or environmental cues, to secure the desired outputs.

Even if this is so, evolutionary psychologists must clearly distinguish between their research agenda and any public policy agenda made on the basis of such research. Public policy recommendations are not strictly part of the evolutionary psychology research programme. To move from hypothesis generation and explanation to public policy is to move from science to an admixture of science and political science and philosophy. This must be explicitly signalled.

Furthermore, if one wants to wear a policy hat, if one wants to use the strength of evolutionary psychology work to formulate public policy recommendations, then the work in question had better be well established. The design details of adaptations matter. To be of use to public policy, to inform policy of what interventions are unlikely to work and what are likely to work, we need to know a lot about the design details of postulated adaptations, we need to know the range and type of cues relevant to the purported psychological adaptation and the contingencies, developmental and contextual, that modulate its output. Only mature, established explanations, therefore, should be submitted for policy consideration. And we might not be in that position for some time.

5.3 A New Subversion

If my argument is right, those in the two extreme positions should decamp into a reasonable middle ground, thereby cooling the debate and facilitating the kind of multi-disciplinary activity needed to vindicate adaptationist claims. Nevertheless, there is still one final opportunity to subvert evolutionary psychology. One could concede that evolutionary psychology has been a useful tool. But is it still a useful tool? Perhaps it's antiquated. Perhaps its time to replace it with a newer, shinier tool.

This line of thought has a distinctive pedigree. The tone of sceptics like Buller, Richardson and Kitcher, the wholesale dismissal of evolutionary psychology, can alienate even those who consider themselves critics of evolutionary psychology. Laland (2007: 7), for example, says that when he reads the leading evolutionary psychologists like Steve Pinker, Martin Daly and John Tooby, he is motivated to criticise evolutionary psychology. However, when he read Buller's *Adapting Mind*, another feeling arose, a feeling to defend evolutionary psychology from what he saw as an 'unjust denigration'. According to Laland, the world is neither black nor white. It is grey. And in such a world, can we trust the judgement of those who see only black and white? Laland wisely cannot.

So this is a grey scepticism, one neither harshly black in condemnation nor snow-white in praise. This milder, more moderate scepticism, a scepticism championed by Kevin Laland, Gillian Brown and Johan Bolhuis, recognises evolutionary psychology's achievements but, crucially, contends that evolutionary psychology's tenets are no longer tenable.

Unlike some of the hardened sceptics, those who subscribe to this moderate scepticism recognise that evolutionary psychology has made valuable contributions: 'There are undoubtedly some very fine pieces of work that show genuine promise of being able to decipher the evolved structures of the mind. The best of evolutionary psychology is as rigorous and sophisticated as any research carried out in the general area of human behaviour and evolution' (Laland and Brown, 2011a: 137). This is an important recognition, made by a leading researcher outside the evolutionary psychology camp but firmly within the broader evolutionary behavioural science field. Such a recognition is the kind of thing that should be welcomed and encouraged.

Nevertheless, Laland and Brown argue that evolutionary psychology needs to change, needs to 'modernise'. Laland and Brown have been arguing this for a number of years now. For example, they (2002: 187) claim

> The fact that few evolutionary psychology studies refer to the findings of modern evolutionary biology reinforces the suspicion that evolutionary psychology has become detached from recent developments in evolutionary thinking, which over the last 30 years have increasingly stressed a wide range of processes.

Bolhuis *et al.* (2011) identify three tenets associated with evolutionary psychology that they judge to be no longer tenable.

The first tenet they associate with evolutionary psychology is gradualism, the idea that evolutionary change, particularly with respect to complex adaptations, occurs slowly. To challenge this tenet, they cite some examples of recent selection. However, recall we already challenged this objection in Chapter 2. In a nutshell, evidence of recent selection for simple physiological traits does not subvert the argument that there's been insufficient time for selection for complex psychological traits; not a single recently evolved complex psychological trait has been proposed; even if selection could select for complex psychological traits in a short period of time, the instability of recent social environments problematises their selection; and even if new complex traits have recently been selected for, this doesn't undermine the search for complex traits with longer pedigrees.

The second tenet they associate with evolutionary psychology is universalism, a belief in the species-typicality of psychological adaptations, which they criticise as blinding evolutionary psychologists to behavioural variation:

> EP's [evolutionary psychology's] emphasis on a universal human nature has hindered its exploitation of new opportunities to examine human diversity utilizing evolutionary biology. Contemporary evolution theory makes predictions about behavioural variation within and between populations in traits commonly studied by evolutionary psychologists. For example, sex differences in mate preferences constitute a large proportion of EP research and are generally assumed to exhibit universal patterns [...]; however, sexual selection theory suggests that a number of factors, such as sex-biased mortality, population density, and variation in mate quality, will affect sex roles [...]. A modern EP would make greater use of the theoretical insights of modern evolutionary biology as a source of testable hypotheses (Bolhuis *et al.*, 2011: 2).

Does a belief in the species typicality of psychological adaptations blind evolutionary psychologists to behavioural variation? It shouldn't do. As we saw in Chapter 2, good design demands good flexibility, and in Chapter 3 we saw how developmental and contextual calibrations are key heuristics. And there is no theoretical reason why evolutionary psychologists cannot adopt the insights of modern evolutionary biology to further develop their testable hypotheses. What Bolhuis *et al.* offer, therefore, is a sociological objection.

Bolhuis *et al.* also see evolutionary psychology's subscription to the species typicality of psychological adaptations as blinding researchers to the possibility of genetic variation biasing the configuration of those adaptations. Bolhuis *et al.* (2011: 3) point out that, 'While variation within populations accounts for the bulk of human genetic variation, around 5%–7% of genetic differences can be attributed to variation between populations ... genetic variation could lead to biases in the human cognitive processing between, as well as within, populations'. However, this misreads the situation. It's not that evolutionary psychologists' subscription to species typicality of psychological adaptations blinding them to the possibility of genetic variation within and between populations biasing the calibration of our psychological adaptations. Rather, the issue is a pragmatic one: evolutionary psychologists begin hypotheses relatively simple, at the level of species-typical design, and over time they can make their hypotheses more complex—such as by incorporating relevant research from behavioural genetics. One has to start somewhere. The dialectical movement from simplicity to complexity, which I identified in Chapter 3, is how hypotheses often progress.

Furthermore, Bolhuis *et al.* blame evolutionary psychology's universalism as leading evolutionary psychologists to running experiments primarily with WEIRDs (Western Educated Industrial Rich and Democratic): 'the notion of universalism has led to the view that undergraduates at Western universities constitute a representative sample of human nature, a view that has been subject to criticism from anthropologists and psychologists' (Bolhuis *et al.*, 2011: 2). However, running experiments with WEIRD undergraduates is merely a convenience, a first step, not a substitute for vigorous cross-cultural research. And as we saw in the previous chapter, evolutionary psychologists are at the forefront of cross-cultural research in psychology.

The third tenet they associate with evolutionary psychology is massive modularity. They claim that massive modularity is not supported by the neuroscientific evidence as it represents 'an unassailable case for the existence of domain-general mechanisms' (Bolhuis *et al.* 2011: 3). However, recall from Chapter 1 that we can distinguish between two senses of massive modularity: a moderate form, which claims that the number of psychological adaptations is substantially greater than has been traditionally believed, and the stronger form, which claims that there are no general purpose mechanisms. If it is indeed the case that there is good evidence for the existence of domain-general mechanisms, then this only speaks against the strong form of massive modularity, not the moderate form. 'Both domain-specific and domain-general mechanisms are compatible with evolutionary theory, and their relative importance in human information processing will only be revealed through careful experimentation, leading to a greater understanding of how the brain works' (*ibid.*, 2011: 3). Agreed. And I'm not entirely sure who would disagree.

So none of the three objections they raise discredits evolutionary psychology, either the top-down, paradigmatic form championed by its main proponents, or the streamlined form I have been championing in this book. More troubling, I find, are the conclusions they draw. After claiming (incorrectly) that the objections they've raised call into question the key tenets of evolutionary psychology, Bolhuis *et al.* (2011: 6) claim that evolutionary psychology should 'reconsider its basic tenets'. 'EP should change its daily practice' (*ibid.*: 6). Evolutionary psychology needs to modernise: 'A modern EP would embrace a broader, more open, and multidisciplinary theoretical framework, drawing on, rather than being isolated from, the full repertoire of knowledge and tools available in adjacent disciplines' (*ibid.*: 6). In short, they call for a 'new science of the evolution of the mind' (*ibid.*: 3).

This judgement, that evolutionary psychology needs to change, needs to modernise, is somewhat ambiguous. Two interpretations are possible. The first interpretation is that evolutionary psychologists need to become more aware, at least to some degree, of what is going on in adjacent evolutionary disciplines, and to look for ways to draw on the tools and insights of these adjacent disciplines to make their own work more sophisticated. The second interpretation is that evolutionary psychologists need to focus less on adaptive problems and adaptive solutions and to instead use more 'modern' ideas, like gene–culture coevolution and niche construction.

The basic thought underlying the first claim is reasonable enough (though one can dispute the characterisation of evolutionary psychologists being isolated from other evolutionary disciplines—and we will, in a moment). Who could disagree with the contention that researchers should know what adjacent disciplines are up to, and to borrow tools and ideas where relevant? The second claim, however, is deeply subversive, and I strongly reject it.

I don't know which interpretation the authors intend. Certainly, it could be read as just a sociological 'know thy neighbour'. But the way the authors position themselves, as claiming that evolutionary psychology is 'required' to change its 'daily practice', that there should be a 'new science' of evolutionary psychology, suggests the second interpretation. The ambiguity alone is subversive. Let's run through each interpretation.

As I said, it's difficult to disagree with the first possible claim. Disciplines necessarily become specialised. There will always be a division of labour. As I argued in the previous chapter, it's not credible to suppose evolutionary psychologists can do all the work that the hard sceptics expect of them. They cannot be jacks of all trades. Nevertheless, it's certainly possible and indeed beneficial to be aware of what's going on in these other disciplines. There is much one could learn by exposing oneself to alternative methods and ideas.

So far, so good. However, this is true of all disciplines. Name me a discipline where it cannot be said that it would benefit from interdisciplinary awareness and cross-fertilisation of methods? And yet this message is especially pressed against evolutionary psychologists, as if they especially need schooling on this point.

Laland and Brown (2002: 187) claim that it's a 'fact' that 'few evolutionary psychology studies refer to the findings of modern evolutionary biology'. But this 'fact' is nothing of the sort. Machery and Cohen (2012) note that the charge of ignoring much of the biological sciences is sometimes pressed more strongly against evolutionary psychology than other research traditions in the evolutionary behavioural sciences. Machery and Cohen (2012) examined the claim that evolutionary psychologists and other evolutionary behavioural scientists are ignorant of the biological sciences. Using quantitative citation analysis, they found the claim to be false. According to Machery and Cohen, evolutionary behavioural scientists, including evolutionary psychologists, are not ignorant of the biological sciences—the proportion of citations of biological science in evolutionary psychology work and other evolutionary behavioural traditions is proportional to biology's scientific output. 'Furthermore, focusing on evolutionary biology in particular instead of the biological sciences in general, it is also not the case that evolutionary behavioral scientists ignore evolutionary biology' (*ibid.*: 217).

So the characterisation of evolutionary psychology as an island stranded in time and isolated from many recent currents of evolutionary biology is evidentially unwarranted. Hence, there is no need for the first possible claim to be especially raised and pressed against evolutionary psychology. Indeed, when characterising a research programme, one must resist the temptation to subscribe to supposed 'truths' about that programme that are only truths in virtue of frequent repetition. You've probably heard from one source or another the view that, 'Perhaps too much research in the field [of evolutionary psychology] is a documentation of what is already known, accompanied by a post hoc evolutionary spin and a snappy press release' (Laland and Brown, 2011a: 137) But Carruthers correctly notes that 'There is now an *immense* body of scientific work in the evolutionary psychology tradition, of most of which the philosophical critics are simply ignorant' (2006: 37, original emphasis). And in previous chapters we found plenty of instances of evolutionary psychology going beyond the documentation of what is already known.

Turning now to the second possible claim, the message here appears to run something like this. Evolutionary psychology's achievements are genuine and permanent additions to knowledge. We can never go back to a state of innocence about the role of selection in shaping our psychology. But like all movements, evolutionary psychology was bound to have a limited life of novelty and vitality. It's become boringly predictable to many people who were once sympathetic to it. There's more to evolutionary theory than adaptationism. Evolutionary theory has evolved.

I can agree with the motivation of the first possible claim, even if it's based on an evidentially unwarranted characterisation. But I profoundly disagree with this stronger claim. Beyond becoming more informed of their peers' work, evolutionary psychologists should not change their daily practice. They should remain doing precisely what they're doing, applying the framework I have been articulating and vindicating in this book. I have absolutely no objection to other methods and ideas being pursued concurrently in the evolutionary behavioural sciences, in fact three cheers for pluralism, but I strongly reject the contention that evolutionary psychologists themselves need to down their tools in favour of other tools and ideas.

The claim that evolutionary psychology needs to change its daily practice in this stronger sense is deeply subversive. Although the hard scepticism and moderate scepticism have different approaches towards evolutionary psychology—wholesale dismissal versus a qualified judgement as to its success—they nevertheless arrive at the same endpoint. Just as the ceaseless evidential requirements of the hard sceptics threaten to deprive us of valuable discoveries, so too does the admonition to change daily practice.

I contend that the basic mistake of the hard sceptics is overlook the heuristics. The milder sceptics recognise the heuristics, but their basic mistake is not weighing those heuristics as importantly as they deserve to be. Laland and Brown (2011b: 358–9, 361) write,

> Comparative statistical and phylogenetic analyses could be deployed to investigate the factors that explain variation in human psychological attributes, but Evolutionary Psychologists generally do not deploy these methods; we suspect because they believe psychological mechanisms are universals. Deprived of these methods, Evolutionary Psychologists are left with little that they can productively do with evolutionary theory apart from using it to generate hypotheses ... evolutionary biology has far more to offer than hypothesis-generation'.

Unlike the hardened sceptics, Laland and Brown get the basic picture of evolutionary psychology right: the generation of testable hypotheses. The importance of this should not be missed: it shows that what evolutionary psychology is really about can be understood by those outside the programme. Yet Laland and Brown appear not to value this activity as highly as this book does. Evolutionary psychologists, according to this view, should be doing other things. They should be doing comparative statistical and phylogenetic analyses, as well as deploying gene–culture co-evolutionary analyses.

This, I believe, underappreciates the specific function evolutionary psychology has in psychology. Hypothesis generation is precisely its strength. As we've seen, although the name 'evolutionary psychology' suggests the programme should be doing a host of other activities, its actual role is one of discovery, of finding new design features of extant mechanisms and new mechanisms entirely. Moderate sceptics don't weigh this heuristic as importantly as it deserves to be weighed. Perhaps this is because they do not sufficiently take into consideration the reality of the unconscious and the problems that creates for discovery. I contend that the evolutionary and behavioural sciences need a programme dedicated to discovering selected psychological processes because, if they exist, they will be largely unconscious to us. Psychological adaptationist thinking provides a non-accidental way of discovering them—perhaps the only non-accidental way of discovering them. This function is very important. If evolutionary psychologists did other activities, this unique function, this important function, would be reduced or lost.

5.4 The Need for Evolutionary Psychology as a Heuristic Project

While the idea of the unconscious may seem commonplace, it is not, in fact, at the heart of many social sciences disciplines, nor is it a common assumption in the daily experience of most people. Consciousness is limited to a narrow sequence of thoughts and events. Few realise the scale of physiological and psychological processes operating below the threshold of consciousness. And even if we recognise and consent to the idea of an unconscious it's difficult to feel this reality in our bones, so to speak.

Many physiological processes happen at once: breathing, heart beating, food digestion, hair growth, body temperature re-calibration. All these physiological processes can occur, and often do occur, without thinking, without being conscious of these processes, though sometimes, of course, we tumble into awareness of breathing or the heart beating. Even the act of seeing, which seems so simple, belies a multitude of complex processes, which ordinary experience alone would fail to inform us about. The same is obviously true of psychological processes. Every moment, many psychological processes are in operation—below the threshold of conscious awareness. Selection, it seems, doesn't care much for conscious awareness. It's rather stingy with it. Consciousness is simply not needed for the tens or hundreds of thousands of concurrent physiological and psychological processes operating in each of us this very moment.

Evolutionary psychologists investigate processes that are largely unconscious to us. Natural selection might have selected for a wide variety of adaptations—but such adaptations do not require conscious awareness of their operation. We might find symmetrical faces attractive, but we need not be consciously aware of this, and certainly need not be aware that this because such cues correlate with health. Given the range of adaptive problems our ancestors likely faced, potentially there is much awaiting discovery. Evolutionary psychology offers a torchlight on seeing psychological adaptations. As Tooby and Cosmides (1992: 67) put it, 'The tools of evolutionary functional analysis function as an organ of perception, bringing the blurry world of human psychological and behavioral phenomena into sharp focus and allowing one to discern the formerly obscured level of our richly organized species-typical functional architecture'.

Take the male adaptation for cue detection of ovulation. Who would have even thought of that without adaptationist thinking? Perhaps one could have stumbled on it by accident, as a lucky conjecture, thought or dream. Perhaps a random guessing game could have done the job. But it seems only an adaptationist perspective could deliver this proposal in a methodological way; non-adaptationist perspectives would simply be blind to the possibility. Or take the impact of olfactory cues on attraction. Before evolutionary psychology came along, despite existing research on olfactory communication in animals, little research had been done on the impact of olfactory cues in human attraction. Non-evolutionary psychologists were simply blind to the possibility of olfactory cues having a role in human attraction, despite the huge market for fragrances and perfumes. Without a programme dedicated to search-lighting selected psychological processes, we would remain blind to these largely unconscious psychological adaptations. We would have little hope of uncovering them.

That many of the processes evolutionary psychologists investigate are unconscious also accounts for why their hypotheses are rewarded with the predictable harvest of incredulousness, ridicule and dismissal. When I first read about an adaptation for ovulation detection, I, too, originally found the idea to be outlandish. How on earth could that be? I can't detect ovulatory cues. I have no clue whether my partner is ovulating or not. And I'm not consciously thinking that when I meet women. I initially believed that this is the kind of thing associated with evolutionary psychology that needed sidelining. But only when I truly appreciated how unconscious many processes are, how selection can engineer truly marvellous mechanisms for competition, survival and reproduction, how information-rich cues can be, did I get the 'light bulb' moment when I believe I really understood evolutionary psychology. Suddenly it all made sense. I now take ovulation detection not as an embarrassing example of the programme, but as a flagship example of the programme. And if I, a person sympathetic to evolutionary psychology, could have initially found such an idea difficult to accept, how much more so for those who hold that evolutionary psychology is 'wrong in every detail'?

Furthermore, if we have a naive or immature understanding of the mind, if we think conscious experience represents the near totality of psychology, rather than the tip of a very deep iceberg, then reading evolutionary psychology research is likely to clash with our personal narratives about what we do and why. It's very common to carry along some very strong notions of who we are or the kind of person we'd like to be: 'I'm this sort of person. I do this for such and such reasons'. That's not helpful, because personal identity is very nebulous, based on all kinds of information and misinformation. It often gets in the way of a scientific understanding of our behaviour.

To paraphrase Jung, knowledge and experience of the unconscious is a defeat for the ego. Those who pride themselves on their personal identity narratives, of seeing themselves as islands of sovereignty, might be offended to learn of selected processes working independently of their socially constructed identities. Hence, perhaps, the common accusation we hear from the more noisy, the more layman commentators that evolutionary psychology is 'dehumanising'. Perhaps it requires a certain maturity of self-awareness, a certain experience that the ego is a satellite, not a star, to be in a position to be receptive to evolutionary psychology thought.

So to answer the question disobligingly asked by the hard sceptics—'Why generate adaptationist hypotheses in psychology?'—and to answer the question asked by the moderate sceptics—'Why only generate adaptationist hypotheses? Why not do other things?'—we can reply that the large number of unconscious processes demands a specialised programme of discovery, that the possibility of discovering unconscious adapted psychological processes is the central idea that animates evolutionary psychology. To suggest, therefore, that evolutionary psychologists should be doing things other than what they're doing is to misunderstand or underappreciate their unique role in the evolutionary and behavioural sciences.

The real issue is not whether evolutionary psychology is outdated, but whether it's exhausted, whether it's just recycling old points. And there are absolutely no indications that it is. It shows no sign of slackening. If anything, it's still early days for evolutionary psychology. I venture the prediction that future evolutionary psychology work will, *inter alia*, increasingly focus on the subtle ways in which early conditions can modulate behavioural tendencies later in life, apply itself to new topics and focus on possible mechanisms that exist in virtue of evolutionary arm races. I further venture that, once the debate is depolarised, once all sides of the debate have a clearer understanding of the role of evolutionary psychology in the evolutionary and behavioural sciences, genuine multidisciplinary cooperation can thrive, offering rich rewards. That's not to say evolutionary psychology doesn't have a shelf life. A torchlight will eventually run out of battery. In common with all programmes, one day its style of reasoning will reach diminishing returns. But it is not this day.

There's a huge privilege given to new ideas. There's been talk in recent years of 'new thinking' in the evolutionary behavioural sciences, indeed of extending and reformulating evolutionary theory itself. While such ideas may sound exciting, whether they can be translated into pragmatic programmes of research is an open question. Psychologists, at least the ones attracted to evolutionary psychology, are strikingly pragmatic. They're interested in what works, in making new discoveries. And the heuristics of evolutionary psychology work effectively for that goal. Perhaps because my own temperament is also pragmatic, I've been able to see virtues in evolutionary psychology that some of my peers, who perhaps have different temperaments, miss. There's also a huge privilege given to formal methods and formal modelling. Economics, for example, has become an overwhelming mathematical subject. We shouldn't devalue informal methods. The evolutionary behavioural sciences can benefit from both formal and informal methods, from mathematical and verbal methods. They complement one another.

Perhaps some of those who are interested in 'newer' ideas and different methods presuppose that because they are interested in such things that therefore all researchers should be. That would be wrong. As Dennett (2011: 483) says in a different context, 'one should always ask whether instead one can have one's cake and eat it too'.

To decide to be uninterested in adaptationist questions is a matter of personal taste. But to insist that these are pointless questions or outdated questions is to deny an important aspect of human psychology and behaviour that warrants scientific study. Evolutionary psychologists have a job to do, and other researchers have other jobs to do. Over time, and in tandem, we have a fair chance of uncovering the tiniest details that constitute unconscious processes. Asking adaptationist questions, asking selectionist and engineering questions about the fit between adaptive problems and adaptive solutions, is an essential part of that collective effort. It's important to find out whether we have psychological adaptations and how they function. And we won't find out without asking. Evolutionary psychologists must therefore keep asking adaptationist questions, without being dismissed as 'idle Darwinisers' or asked to 'modernise'.

5.5 Conclusion

The sceptics' fundamental mistake is to overlook the heuristics of the programme and to instead overwhelmingly focus on the current evidential status of psychological adaptation claims. I submit that the sceptics need to recognise that evolutionary psychology is heuristically driven. The moderate sceptics' basic mistake is to undervalue the heuristics and to thereby advocate that evolutionary psychology should be doing things other than generating testable hypotheses. I submit that the moderate sceptics need to recognise that we really do need a dedicated adaptationist programme of exploration, as we will be largely unconscious of any adapted psychological processes we might have. Prominent evolutionary psychologists' mistake is to obscure or overshadow the heuristics by bundling them with tenets irrelevant to its practice. Given that evolutionary psychology needs multiple disciplines to engage with its findings, needs multiple disciplines to take seriously what the programme has to offer the evolutionary and behavioural sciences, this will prove to be an increasingly costly mistake if not corrected.

Kurzban (2010) says, 'I would argue that perhaps the field's greatest challenge lies less [in] coaxing nature to give up her secrets, and more in communicating the insights from evolutionary psychology to those outside the field'. In common with many evolutionary psychologists, Kurzban is perplexed as to why many people have so badly mistaken evolutionary psychology. In light of our discussions, we should now be less perplexed. People are badly mistaking evolutionary psychology because they've got the basic picture of what evolutionary psychology is about wrong and they've got the basic picture wrong, in part because of the way the programme has been presented and packaged.

Being clear about the role and limits of evolutionary psychology is essential if we are to make the case that evolutionary psychology has a serious and important role to play in the evolutionary and behavioural sciences. If prominent evolutionary psychologists want evolutionary psychology to truly realise its potential of stimulating multidisciplinary activity towards establishing the multiple lines of evidence needed to fully develop and vindicate adaptationist explanations, then they must clearly and efficiently communicate its research goals to those in adjacent disciplines, in a streamlined way, not in an inflated way.

I end this chapter with a call to action. Hard sceptics: as evolutionary psychology functions heuristically, focus on that. You're quite right in drawing attention to the incompleteness of evidential portfolios but it's for a wider constituency to fill in the details collectively over time. Prominent evolutionary psychologists: refocus attention on what's valuable about adaptationist thinking in psychology; disassociate tenets not strictly part of the research programme. Moderate sceptics: realise that the evolutionary behavioural sciences would be poorer if evolutionary psychologists changed their daily practice. New thinking and developments in evolutionary theory might indeed speak in favour of new research programmes in the evolutionary behavioural sciences but that does not mean evolutionary psychologists should change their daily practice.

If the various parties endorsed these points, the polarisation can end, and the fruits of genuine multidisciplinary research on purported psychological adaptations, as outlined in the previous chapter, should begin to grow.

Conclusions

We evolved through natural selection. Selection has crafted and refined our physiology into multiple functionally specialised mechanisms. Might it also have crafted and refined our psychology in a similar way? There is no *a priori* guarantee that it has. But it should be clear that such a notion is a real contender. And if that is indeed the case, and there are no reasonable grounds to dismiss this, then might not hypothesising about adaptive problems and adaptive solutions yield new insights? As imperfect as the method is, might it not yield new and permanent editions of knowledge?

This thought was developed and launched into a programme of research in the early 1990s by pioneering thinkers and continues to grow in both research output and research affiliates. They also packaged this thought into an overly ambitious paradigm. In doing so and in baptising the programme as 'evolutionary psychology', I believe they made a fundamental strategic mistake—which purchased visibility with the public at the expense of respectability with peers, a mistake that has disabled understanding and disfigured the debate to this very day.

Nevertheless, I believe that once attention is decisively shifted from evolutionary psychology as a naive explanatory project to a cautious exploratory project, once it is unbundled as a paradigm and reframed as a streamlined research programme, I believe any reasonable person, no matter her conception of evolution, no matter her conception of the mind, no matter her outlook on the ultimate empirical prospects and bets of adaptationist thinking in psychology, will consent to the legitimacy, to the value and to the role of evolutionary psychology in the evolutionary behavioural sciences.

It is precisely this shift in attention, this shift in focus and in emphasis, the need for it, how it can be done and the ramifications of doing so, that this book has sought to explicate and to champion. No doubt various particular points can and will be disputed. But I believe the overall shift is as secure as it is needed.

Evolutionary psychology is a hypothesis-driven empirical science. I've argued that evolutionary psychology can:

H1 Hypothesise unknown design features of extant psychological mechanisms, leading to novel predictions of psychological phenomena
H2 Hypothesise unknown psychological mechanisms, leading to novel predictions of psychological phenomena
H3 Hypothesise traits as by-products of psychological mechanisms, leading to novel predictions of psychological phenomena
H4 Stimulate multidisciplinary research activities, leading to more sophisticated hypotheses and more complete ultimate and proximate explanations

As evolutionary psychology is in the business of generating and testing hypotheses about psychological mechanisms, it is not only eminently reasonable, but should also be welcomed by all.

Suppose we observe trait *T*. It's developmentally reliable and robust. It can be observed across cultures. It's a suitable candidate for adaptationist hypothesising—not that it must be an adaptation, but that it potentially could be and to an extent that merits further investigation. Accordingly, we deploy an adaptationist analysis. We can ask whether trait *T* has a function. If so, how has natural selection designed that mechanism to serve that function? Posing these questions can lead to new insights. We can think in new ways about how the mechanism develops and operates, such as developmental and environmental contexts that calibrate the mechanism, the cues that activate the mechanism, the strength and variety of outputs, and so on. And, crucially, we can articulate these to the point of testability.

Potentially much understanding is to be won by hypothesising such traits as adaptations. It's commonplace that often the most difficult thing to see is right in front of one's nose. We need theories and hypotheses to guide us into identifying patterns. Evolutionary psychology can guide researchers into identifying new behavioural patterns, new demographic patterns, new facts—facts that do not readily lend themselves to discovery by non-adaptationist perspectives. And potentially groundbreaking, reflecting on adaptive problems and scenarios holds out the promise of discovering entire mechanisms we would unlikely find by other methods.

Initial evolutionary psychology hypotheses aim, or should aim, not for the last evolutionary word on a given phenomenon, but the first. They are in constant adjustment—both with the research programme's own findings and findings from adjacent research programmes and disciplines. If this is done, this should generate sophisticated hypotheses, as well as generate progressive increments to our understanding of psychological and social phenomena. In virtue of its successful novel predictions, evolutionary psychology can help stimulate multidisciplinary activity towards establishing the multiple lines of evidence needed to fully vindicate adaptationist explanations. The forthcoming PsychTable, which promises to codify and evaluate the range and strength of evidence for purported psychological adaptations, should help foster the interdisciplinary cooperation needed to vigorously vindicate psychological adaptation claims.

Given the range of phenomena that psychological adaptations, in various degrees, can influence, and given that psychological adaptations can generate a multitude of surprising by-products, evolutionary psychology has a wide remit. Finally, as a general mode of scientific procedure, we are inclined to accept hypotheses insofar as they make successful predictions of new data. At the very least, evolutionary psychology is capable of meeting this standard and has done so in multiple cases.

This is the positive case for evolutionary psychology. The streamlined case. What I have been championing and what I believe best represents evolutionary psychology practice. There should be little to find objectionable about it.

The moderate sceptic claims evolutionary psychologists should change their daily practice. There's more to evolutionary theory than adaptationism. That's true, but this fails to appreciate the importance of such thinking in the discovery of unconscious processes. The real issue is not whether evolutionary psychology is outdated, but whether it's exhausted, whether it's just recycling old points. And there are no indications that it is.

The strong sceptic focuses on the evidential status of evolutionary psychology explanations. She sets out the tough standards complete adaptationist explanations should meet, notes the shortfall and then dismisses evolutionary psychology. But evolutionary psychologists do not have to meet the heavy evidential requirements identified by Richardson and others in order to pursue legitimately adaptationist theorising in psychology. To insist these heavy requirements be met at the outset is dangerous. And to insist or to imply that evolutionary psychology alone meet these heavy evidential demands is disingenuous. Yes, evolutionary psychology cannot possibly meet the evidential demands—but evolutionary psychology doesn't need to do this alone. And there's no reason to suppose the evolutionary and behavioural sciences cannot do so collectively.

If evolutionary psychology is to cease being dismissed, its champions need to shift attention to its heuristics, to its rich possibilities of discovery across the spectrum of social phenomena. These heuristics represent a unique opportunity to discover hitherto unknown design features of known traits and to even discover entire traits hitherto unknown, to make genuine and permanent editions to knowledge. That's precious—too precious to be overshadowed by theoretical excess. Too precious to be knocked about by jibes about just so stories.

References

Alcock, J. and C. Crawford (2008) 'Evolutionary Questions for Evolutionary Psychologists', in C. Crawford and D. Krebs (eds) *Foundations of Evolutionary Psychology*, pp. 25–46 (New York: Lawrence Erlbaum Associates).

Atran, S. (2002) *In Gods We Trust: The Evolutionary Landscape of Religion* (Oxford: Oxford University Press).

Balachandran, N. and D. J. Glass (2012) 'PsychTable.org: The Taxonomy of Human Evolved Psychological Adaptations', *Evolution: Education and Outreach* 5:312–20.

Barkow, J. H. (2006) 'Introduction: Sometimes the Bus Does Wait', in J. H. Barkow (ed.) *Missing the Revolution: Darwinism for Social Scientists*, pp. 3–60 (Oxford: Oxford University Press).

Barkow, J. H. (2009) 'Steps Toward Convergence: Evolutionary Psychology's Saga Continues', *Proceedings of the National Academy of Sciences of the United States of America* 106:14743–4.

Barkow, J., L. Cosmides and J. Tooby (eds) (1992) *The Adapted Mind: Evolutionary Psychology and the Generation of Culture* (New York: Oxford University Press).

Barrett, D. (2010) *Supernormal Stimuli: How Primal Urges Overran Their Evolutionary Purpose* (New York: WW. Norton and Company).

Barrett, H. C. (2006) 'Modularity and Design Reincarnation', in P. Carruthers, S. Laurence and S. Stich (eds) *The Innate Mind Volume 2: Culture and Cognition*, pp. 199–217 (New York: Oxford University Press).

Barrett, H. C. (2007a) 'Development as the Target of Evolution: A Computational Approach to Developmental Systems', in S. Gangestad and J. Simpson (eds) *The Evolution of Mind: Fundamental Questions and Controversies*, pp. 186–92 (New York: Guilford).

Barrett, H. C. (2007b) 'Modules in the Flesh', in S. Gangestad and J. Simpson (eds) *The Evolution of Mind: Fundamental Questions and Controversies*, pp. 161–8 (New York: Guilford).

Barrett, H. C. (2011) 'The Wrong Kind of Wrong: A Review of What Darwin Got Wrong by Jerry Fodor and Massimo Piatelli-Palmarini', *Evolution and Human Behavior* 32:76–8.

Barrett, J. L. (2004) *Why Would Anyone Believe in God?* (Walnut Creek, CA: AltaMira Press).

Baumeister, R. F., K. D. Vohs and D. M. Tice (2007) 'The Strength Model of Self-Control', *Current Directions in Psychological Science* 16:351–5.

Becker, D. V., B. J. Sagarin, R. E. Guadagno, A. Millevoi and L. D. Nicastle (2004) 'When the Sexes Need not Differ: Emotional Responses to the Sexual and Emotional Aspects of Infidelity', *Personal Relationships* 11:529–38.

Belsky, J., L. Steinberg and P. Draper (1991) 'Childhood Experience, Interpersonal Development, and Reproductive Strategy: An Evolutionary Theory of Socialization', *Child Development* 62:647–70.

Berna, F., P. Goldberg, L. K. Horwitz, J. Brink, S. Holt, M. Bamford and M. Chazan (2012) 'Microstratigraphic Evidence of in situ Fire in the Acheulean

Strata of Wonderwerk Cave, Northern Cape Province, South Africa', *Proceedings of the National Academy of Sciences of the United States of America* 109:E1215–20.

Bjorklund, D. F. and J. M. Bering (2002) 'The Evolved Child: Applying Evolutionary Developmental Psychology to Modern Schooling', *Learning and Individual Differences* 12:1–27.

Bolhuis, J. J., G. R. Brown, R. C. Richardson and K. N. Laland (2011) 'Darwin in Mind: New Opportunities for Evolutionary Psychology', *PLoS Biology* 9:1–8.

Borrello, M. E. (2005) 'The Rise, Fall and Resurrection of Group Selection', *Endeavour* 29:43–7.

Boyer, P. (2002) *Religion Explained: The Human Instincts that Fashion Gods, Spirits and Ancestors* (London: Vintage).

Brandon, R. N. (1990) *Adaptation and Environment* (Princeton, NJ: Princeton University Press).

Brown, W. M., L. Cronk, K. Grochow, A. Jacobson, C. Karen Liu, Z. Popović and R. Trivers (2005) 'Dance Reveals Symmetry Especially in Young Men', *Nature* 438:1148–50.

Brown, G. R., T. E. Dickins, R. Sear and K. N. Laland (2011) 'Evolutionary Accounts of Human Behavioural Diversity', *Philosophical Transactions of the Royal Society B: Biological Sciences* 366:313–24.

Bryant, G. A. and M. G. Haselton (2009) 'Vocal Cues of Ovulation in Human Females', *Biology Letters* 5:12–15.

Bulbulia, J. (2004) 'The Cognitive and Evolutionary Psychology of Religion', *Biology and Philosophy* 19:655–86.

Buller, D. J. (2005) *Adapting Minds* (Cambridge, MA: MIT Press).

Buller, D. J. and V. G. Hardcastle (2000) 'Evolutionary Psychology, Meet Developmental Neurobiology: Against Promiscuous Modularity', *Brain and Mind* 1:307–25.

Buss, D. M. (1995) 'Evolutionary Psychology: A New Paradigm for Psychological Science', *Psychological Inquiry* 6:1–30

Buss, D. M. (2005a), 'Introduction: The Emergence of Evolutionary Psychology', in D. M. Buss (ed.) *The Handbook of Evolutionary Psychology*, pp. xxiii–xxv (Hoboken, NJ: John Wiley & Sons).

Buss, D. M. (2005b), 'Mating', in D. M. Buss (ed.) *The Handbook of Evolutionary Psychology*, pp. 251–4 (Hoboken, NJ: John Wiley & Sons).

Buss, D. M. (2005c) 'Foundations of Evolutionary Psychology', in D. M. Buss (ed.) *The Handbook of Evolutionary Psychology*, pp. 1–3 (Hoboken, NJ: John Wiley & Sons).

Buss, D. M. (2008) *Evolutionary Psychology: The New Science of the Mind*, 3rd ed. (Boston, MA: Allyn & Bacon).

Buss, D. M. and T. K. Shackelford (1997) 'From Vigilance to Violence: Mate Retention Tactics in Married Couples', *Journal of Personality and Social Psychology* 72:346–61.

Buss, D. M., and H. Greiling (1999) 'Adaptive Individual Differences', *Journal of Personality* 67:209–43.

Buss, D. M. and H. K. Reeve (2003) 'Evolutionary Psychology and Developmental Dynamics', *Psychological Bulletin* 129:848–53.

Buss, D. M. and M. Haselton (2005) 'The Evolution of Jealousy', *Trends in Cognitive Sciences* 9:506–7.

Buss, D. M. and J. D. Duntley (2011) 'The Evolution of Intimate Partner Violence', *Aggression and Violent Behavior* 16:411–19.

Buss, D. M., R. J. Larsen, D. Westen and J. Semmelroth (1992) 'Sex Differences in Jealousy: Evolution, Physiology, and Psychology', *Psychological Science* 3:251–5.

Buss, D. M., M. G. Haselton, T. K. Shackleford, A. L. Bleske and J. C. Wakefield (1998) 'Adaptations, Exaptations, and Spandrels', *American Psychologist* 53:533–48.

Buss, D. M., T. K. Shackelford, L. A. Kirkpatrick, J. C. Choe, M. Hasegawa, T. Hasegawa, *et al.* (1999) 'Jealousy and the Nature of Beliefs About Infidelity: Tests of Competing Hypotheses About Sex Differences in the United States, Korea, and Japan', *Personal Relationships* 6:125–50.

Caldwell-Harris, C. L., C. F. Murphy, T. Velazquez and P. McNamara (2011), 'Religious Belief Systems of Persons with High Functioning Autism', presented at the Annual Meeting of the Cognitive Science Society, Boston, MA, available at: http://csjarchive.cogsci.rpi.edu/proceedings/2011/papers/0782/paper0782.pdf (accessed 22 September 2012).

Callender, C. (2012) 'Time Lord' (interview with Richard Marshall), available at: http://www.3ammagazine.com/3am/time-lord/ (accessed 4 July 2012).

Cameron, N., E. Fish and M. J. Meaney (2004) 'Variations in Maternal Care Influence Mating Preference of Female Rats', presented at the Society for Neuroscience, San Diego, CA, 24 October 2004.

Campbell, A. (2006) 'Feminism and Evolutionary Psychology', in J. H. Barkow (ed.) *Missing the Revolution: Darwinism for Social Scientists*, pp. 63–100 (Oxford: Oxford University Press).

Caporael, L. and M. Brewer (1990) 'We ARE Darwinians, and this is what the Fuss is all About', *Motivation and Emotion* 14:287–293.

Carruthers, P. (2006) *The Architecture of the Mind* (Oxford: Oxford University Press).

Carruthers, P. (2008) 'On Fodor-Fixation, Flexibility, and Human Uniqueness: A Reply to Cowie, Machery, and Wilson', *Mind and Language*, 23:293–303.

Cartwright, J. (2008) *Evolution and Human Behaviour* (New York: Palgrave Macmillan).

Cosmides, L. and J. Tooby (1987) 'From Evolution to Behavior: Evolutionary Psychology as the Missing Link', in J. Dupré (ed.) *The Latest on the Best: Essays on Evolution and Optimality*, pp. 276–306 (Cambridge, MA: MIT Press).

Cosmides, L. and J. Tooby (1992) 'Cognitive Adaptations for Social Exchange', in J. H. Barklow, L. Cosmides and J. Tooby (eds) *The Adapted Mind: Evolutionary Psychology and the Generation of Culture*, pp. 163–228 (New York: Oxford University Press).

Cosmides, L. and J. Tooby (1994) 'Origins of Domain Specificity: The Evolution of Functional Organization', in L. Hirschfeld and S. Gelman (eds) *Mapping the Mind: Domain Specificity in Cognition and Culture* (New York: Cambridge University Press).

Cosmides, L. and J. Tooby (1995) 'From Evolution to Adaptations to Behavior: Toward an Integrated Evolutionary Psychology', in R. Wong (ed.) *Biological Perspectives on Motivated Activities*, pp. 11–74 (Norwood, NJ: Ablex).

Cosmides, L. and J. Tooby (1997a) 'Evolutionary Psychology: A Primer', available at: http://www.psych.ucsb.edu/research/cep/primer.html (accessed 14 December 2011).

Cosmides, L. and J. Tooby (1997b) 'Letter to the Editor of *The New York Review of Books* on Stephen Jay Gould's "Darwinian Fundamentalism" (June 12, 1997)', available at: http://cogweb.ucla.edu/Debate/CEP_Gould.html (accessed 22 January 2010).

Cosmides, L. and J. Tooby (2000), 'Evolutionary Psychology and the Emotions', in M. Lewis and J. M. Haviland-Jones (eds) *Handbook of Emotions*, 2nd ed., pp. 91–115 (New York: Guilford).

Cosmides, L., and J. Tooby (2006) 'Evolutionary Psychology, Moral Heuristics, and the Law', in G. Gigerenzer and C. Engel (eds), *Heuristics and the Law*, pp. 175–205 (Cambridge, MA: MIT Press).

Cosmides, L., J. Tooby and R. Kurzban (2003) 'Perceptions of Race', *Trends in Cognitive Sciences* 7:173–9.

Cosmides, L., J. Tooby, L. Fiddick and G. Bryant (2005) 'Detecting Cheaters', *Trends in Cognitive Sciences* 9:505–6.

Coyne, J. A. (2009) *Why Evolution is True* (Oxford: Oxford University Press).

Crawford, C. and C. Salmon (eds) (2004) *Evolutionary Psychology, Public Policy, and Personal Decisions* (Mahwah, NJ: Lawrence Erlbaum).

Csibra, G. and G. Gergely (2009) 'Natural Pedagogy', *Trends in Cognitive Science* 13:148–53.

Culver, M. (2009) 'Lessons and Insights from Evolution, Taxonomy and Conservation Genetics', in M. Hornocker and S. Negri (eds) *Cougar: Ecology and Conservation*, pp. 27–40 (Chicago, IL: University of Chicago Press).

Curtis, V., R. Aunger and T. Rabie (2004) 'Evidence that Disgust Evolved to Protect from Risk of Disease', *Proceedings of the Royal Society B: Biological Sciences* 271(Suppl. 4):S131–3.

Daly, M. and M. Wilson (1980), 'Discriminative Parental Solicitude: A Biological Perspective', *The Journal of Marriage and Family* 42:277–88.

Daly, M. and M. Wilson (1994) 'Some Differential Attributes of Lethal Assaults on Small Children by Stepfathers Versus Genetic Fathers', *Ethology and Sociobiology* 15:207–17.

Daly, M. and M. Wilson (1996) 'Evolutionary Psychology and Marital Conflict: The Relevance of Stepchildren', in D. M. Buss and N. Malamuth (eds) *Sex, Power, Conflict: Feminist and Evolutionary Perspectives*, pp. 9–28 (New York: Oxford University Press).

Daly, M. and M. Wilson (1999), 'Human Evolutionary Psychology and Animal Behaviour', *Animal Behaviour* 57:509–19.

Daly, M. and M. Wilson (2001) 'Risk-taking, Intrasexual Competition, and Homicide', *Nebraska Symposium on Motivation* 47:1–36.

Daly, M. and M. Wilson (2005a) 'Human Behavior as Animal Behavior' in J. J. Bolhuis and L. A. Giraldeau (eds) *Behavior of Animals: Mechanisms, Function, and Evolution*, pp. 393–408 (Oxford: Blackwell Publishing).

Daly, M. and M. Wilson (2005b) 'The "Cinderella Effect" is no Fairy Tale', *Trends in Cognitive Sciences* 9:507–8.

Daly, M. and M. Wilson (2007) 'Is the "Cinderella Effect" Controversial? A Case Study of Evolution-minded Research and Critiques Thereof', in C. Crawford and D. Krebs (eds) *Foundations of Evolutionary Psychology*, pp. 383–400 (New York: Lawrence Erlbaum Associates).

Daly, M., M. Wilson and S. J. Weghorst (1982) 'Male Sexual Jealousy', *Ethology and Sociobiology* 3:11–27.

D'Arms, J., R. Batterman and K. Gorny (1998) 'Game Theoretic Explanations and the Evolution of Justice', *Philosophy of Science* 65:76–102.

Darwin, C. (1859) *On the Origin of Species* (London: John Murray).

Davies, N. B., J. R. Krebs and S. A. West (2012) *An Introduction to Behavioural Ecology*, 4th ed. (Oxford: Wiley-Blackwell).

Dawkins, R. (1976) *The Selfish Gene* (Oxford: Oxford University Press).

Dawkins, R. (2005) 'Afterword', in D. M. Buss (ed.) *The Handbook of Evolutionary Psychology*, pp. 975–80 (Hoboken, NJ: John Wiley & Sons).

Dawkins, R. (2006) *The God Delusion* (London: Bantam Press).

DeBruine, L. M. (2009) 'Beyond "Just-so Stories"', *The Psychologist* 22:930–2.

Delton, A. W., T. E. Robertson and D. T. Kenrick (2006) 'The Mating Game isn't Over: A Reply to Buller's Critique of the Evolutionary Psychology of Mating', *Evolutionary Psychology* 4:262–73.

Dennett, D. C. (1995) *Darwin's Dangerous Idea* (London: Allen Lane).

Dennett, D. C. (2011) 'Homunculi Rule: Reflections on Darwinian Populations and Natural Selection by Peter Godfrey Smith', *Biology and Philosophy* 26:475–88.

de Sousa Campos, L., E. Otta and J. de Oliveira Siqueira (2002) 'Sex Differences in Mate Selection Strategies: Content Analyses and Responses to Personal Advertisements in Brazil', *Journal of Evolution and Human Behavior* 23:395–406.

de Waal, F. B. M. (2001) 'The Inevitability of Evolutionary Psychology and the Limitations of Adaptationism: Lessons from the Other Primates', *International Journal of Comparative Psychology* 14:25–42.

Dickins T. E. and B. J. A. Dickins (2008) 'Mother Nature's Tolerant Ways: Why Non-genetic Inheritance has Nothing to do with Evolution', *New Ideas in Psychology* 26:41–54.

Duntley, J. D. and D. M. Buss (2008) 'Evolutionary Psychology is a Metatheory for Psychology', *Psychological Inquiry* 19:30–4.

Duntley, J. D. and T. K. Shackelford (2008) 'Darwinian Foundations of Crime and Law', *Aggression and Violent Behavior* 13:373–82.

Duntley, J. D. and T. K. Shackelford (2012) 'Adaptations to Avoid Victimization', *Aggression and Violent Behavior* 17:59–71.

Dupré, J. (2001) *Human Nature and the Limits of Science* (Oxford: Oxford University Press).

Dupré, J. (2003) 'On Human Nature', *Human Affairs* 13:109–22.

Dupré, J. (2012), *Processes of Life: Essays in the Philosophy of Biology* (Oxford: Oxford University Press).

Durham, W. H. (1991) *Coevolution: Genes, Culture and Human Diversity* (Palo Alto, CA: Stanford University Press).

Ermer, E., L. Cosmides and J. Tooby (2007) 'Functional Specialization and the Adaptationist Program', in S. Gangestad and J. Simpson (eds) *The Evolution of Mind: Fundamental Questions and Controversies*, pp. 153–60 (New York: Guilford).

Ernst, Z. (2005) 'A Plea for Asymmetric Games', *Journal of Philosophy* 102:109–25.

Fitzgerald, C. J. and M. B. Whitaker (2010) 'Examining the Acceptance of and Resistance to Evolutionary Psychology', *Evolutionary Psychology* 8:284–96.

Fodor, J. (1983) *The Modularity of Mind* (Cambridge, MA: MIT Press).

Fodor, J. (2000) *The Mind Doesn't Work That Way* (Cambridge, MA: MIT Press).

Frank, R., T. Gilovich and D. Regan (1993) 'Does Studying Economics Inhibit Cooperation?', *Journal of Economic Perspectives* 7:159–71.

Frederick, M. J. (2012) 'Birth Weight Predicts Scores on the ADHD Self-Report Scale and Attitudes Towards Casual Sex in College Men: A Short-Term Life History Strategy?', *Evolutionary Psychology* 10:342–51.

Fuentes, A. (2009) *Evolution of Human Behavior* (New York: Oxford University Press).

Galperin, A. and M. G. Haselton (2012) 'Error Management and the Evolution of Cognitive Bias', in J. P. Forgas, K. Fiedler and C. Sedikedes (eds) *Social Thinking and Interpersonal Behavior*, pp. 45–64 (New York: Psychology Press).

Gangestad, S W. (2000) 'Human Sexual Selection, Good Genes, and Special Design', *Annals of the New York Academy of Sciences* 907:50–61.

Gangestad, S. W. and D. M. Buss (1992) 'Pathogen Prevalence and Human Mate Preferences', *Ethology and Sociobiology* 14:89–96.

Gangestad, S. W. and J. A. Simpson (2007) 'Whither Science of the Evolution of Mind?', in S. Gangestad and J. Simpson (eds) *The Evolution of Mind: Fundamental Questions and Controversies*, pp. 397–437 (New York: Guilford).

Gangestad, S. W. and R. Thornhill (1997) 'Human Sexual Selection and Developmental Stability', in J. A. Simpson and D. T. Kenrick (eds) *Evolutionary Social Psychology*, pp. 169–95 (Mahwah, NJ: Lawrence Erlbaum).

Gangestad, S. W. and R. Thornhill (1998) 'Menstrual Cycle Variation in Women's Preferences for the Scent of Symmetrical Men', *Proceedings of the Royal Society B: Biological Sciences* 265:927–33.

Gangestad, S. W., R. Thornhill and C. E. Garver-Apgar (2005) 'Women's Sexual Interests Across the Ovulatory Cycle Depend on Primary Partner Developmental Instability', *Proceedings of the Royal Society B: Biological Sciences* 272:2023–7.

Gangestad, S. W., C. E. Garver-Apgar, J. A. Simpson and A. J. Cousins (2007) 'Changes in Women's Mate Preferences Across the Ovulatory Cycle', *Journal of Personality and Social Psychology* 92:151–63.

Geary, D. C. (2005) *The Origin of Mind: Evolution of Brain, Cognition, and General Intelligence* (Washington, DC: American Psychological Association).

Geary, D. C. (2012) 'Application of Evolutionary Psychology to Academic Learning', in S. C. Roberts (ed.) *Applied Evolutionary Psychology*, pp. 78–92 (Oxford: Oxford University Press).

Gintis, H. (2007) 'A Framework for the Unification of the Behavioral Sciences', *Behavioral and Brain Sciences* 30:1–61.

Gintis, H. (2009) *The Bounds of Reason: Game Theory and the Unification of the Behavioral Sciences* (Princeton, NJ: Princeton University Press).

Godfrey-Smith, P. (1998) *Complexity and the Function of Mind in Nature* (Cambridge: Cambridge University Press).

Godfrey-Smith, P. (2001) 'Three Kinds of Adaptationism', in S. H. Orzack, and E. Sober (eds) *Adaptationism and Optimality*, pp. 335–57 (New York: Cambridge University Press).

Godfrey-Smith, P. (2009) *Darwinian Populations and Natural Selection* (Oxford: Oxford University Press).

Goetz, A. T., T. K. Shackelford and S. M. Platek (2009) 'Introduction to Evolutionary Psychology', in S. M. Platek and T. K. Shackelford (eds) *Foundations in Evolutionary Cognitive Neuroscience*, pp. 1–21 (Cambridge: Cambridge University Press).

Gopnik, A. (2013) 'Developmental Timing Explains the Woes of Adolescence', in J. Brockman (ed.) *This Explains Everything: Deep, Beautiful, and Elegant Theories of How the World Works*, pp. 320–3 (New York: Harpercollins).
Gould, S. J. (1997) 'Darwinian Fundamentalism', *The New York Review of Books* 44:34–7.
Gould, S. J. (2000) 'More Things in Heaven and Earth', in H. Rose and S. Rose (eds) *Alas Poor Darwin: Arguments Against Evolutionary Psychology*, pp. 101–26 (New York: Harmony Books).
Gray, R. D., M. Heaney and S. Fairhall (2003) 'Evolutionary Psychology and the Challenge of Adaptive Explanation', in K. Sterelny and J. Fitness (eds) *From Mating to Mentality*, pp. 247–68 (London and New York: Psychology Press).
Griffiths, P. E. (2007) 'Evo-Devo Meets the Mind: Towards a Developmental Evolutionary Psychology' in R. Sanson and R. N. Brandon (eds) *Integrating Development and Evolution*, pp. 195–225 (Cambridge: Cambridge University Press).
Griggs, R. A. and J. R. Cox (1982) 'The Elusive Thematic-materials Effect in Wason's Selection Task', *British Journal of Psychology* 73:407–20.
Griskevicius, V., J. M. Ackerman and J. P. Redden (2012) 'Why we Buy: Evolution, Marketing, and Consumer Behaviour', in S. C. Roberts (ed.) *Applied Evolutionary Psychology*, pp. 311–29 (Oxford: Oxford University Press).
Guthrie, S. E. (1993) *Faces in the Clouds: A New Theory of Religion* (Oxford: Oxford University Press).
Hagen, E. H. (2004) 'The Evolutionary Psychology FAQ', available at: http://www.anth.ucsb.edu/projects/human/evpsychfaq.html (accessed 20 November 2010).
Hagen, E. H. (2005) 'Controversial Issues in Evolutionary Psychology', in D. M. Buss (ed.) *The Handbook of Evolutionary Psychology*, pp. 145–76 (Hoboken, N.J.: John Wiley & Sons).
Hamilton, R. (2008) 'The Darwinian Cage: Evolutionary Psychology as Moral Science', *Theory, Culture and Society* 25:105–25.
Hamilton, W. D. (1964) 'The Genetical Evolution of Social Behaviour I and II', *Journal of Theoretical Biology* 7: 1–32.
Hansen, C. H. and R. D. Hansen (1988) 'Finding the Face in the Crowd: An Anger Superiority Effect', *Journal of Personality and Social Psychology* 54: 917–24.
Hardy, K., S. Buckley, M.J. Collins, A. Estalrrich, D. Brothwell, L. Copeland *et al.* (2012) 'Neanderthal Medics? Evidence for Food, Cooking and Medicinal Plants Entrapped in Dental Calculus', *Naturwissenschaften* 99:617–26.
Haselton, M. G. and S. W. Gangestad (2006) 'Conditional Expression of Women's Desires and Men's Mate Guarding Across the Ovulatory Cycle', *Hormones and Behavior* 49:509–18.
Haselton, M. G., M. Mortezaie, E. G. Pillsworth, A. E. Bleske-Recheck and D. A. Frederick (2007) 'Ovulation and Human Female Ornamentation: Near Ovulation, Women Dress to Impress', *Hormones and Behavior* 51:40–5.
Haufe, C. (2008) 'Perverse Engineering', *Philosophy of Science* 75:437–46.
Heyes, C. (2012) 'New Thinking: The Evolution of Human Cognition', *Philosophical Transactions of the Royal Society B: Biological Sciences* 367:2091–6.
Jackson, R. E. and L. K. Cormack (2007), 'Evolved Navigation Theory and the Descent Illusion', *Perception and Psychophysics* 69:353–62.
Jackson, R. E. and L. K. Cormack (2008) 'Evolved Navigation Theory and the Environmental Vertical Illusion', *Evolution and Human Behavior* 29:299–304.

Jones, B. C., A. C. Little, I. S. Penton-Voak, B. P. Tiddeman, D. M. Burt and D. I. Perrett (2001) 'Facial Symmetry and Judgements of Apparent Health: Support for a "Good Genes" Explanation of the Attractiveness–Symmetry Relationship', *Evolution and Human Behavior*, 22:417–29.

Ketelaar, T., B. L. Koenig, D. Gambacorta, I. Dolgov, D. Hor, J. Zarzosa, *et al.* (2012) 'Smiles as Signals of Lower Status in Football Players and Fashion Models: Evidence that Smiles are Associated with Lower Dominance and Lower Prestige', *Evolutionary Psychology*, 10:371–97.

Kitcher, P. (1985) *Vaulting Ambition: Sociobiology and the Quest for Human Nature* (Cambridge, MA: MIT Press).

Kitcher, P. (2009) 'Philip Kitcher', in G. Oftedal, J. K. B. Olsen, P. Rossell and M. S. Norup (eds) *Evolutionary Theory: 5 Questions*, pp. 79–93 (Copenhagen: Automatic Press/VIP).

Kitcher, P. and A. L. Vickers (2003) 'Pop Sociobiology Reborn: The Evolutionary Psychology of Sex and Violence', in P. Kitcher (ed.) *In Mendel's Mirror: Philosophical Reflections on Biology*, pp. 333–55 (Oxford: Oxford University Press).

Kohn, M. (2008) 'Darwin 200: The Needs of the Many', *Nature* 456:296–9.

Kramer, R. S. S., J. King and R. Ward (2011) 'Identifying Personality from the Static, Nonexpressive Face in Humans and Chimpanzees: Evidence of a Shared System for Signaling Personality', *Evolution and Human Behavior* 32:179–185.

Kramer, R. S. S. and R. Ward (2012) 'Cues to Personality and Health in the Facial Appearance of Chimpanzees (*Pan troglodytes*)', *Evolutionary Psychology* 10:320–37.

Krebs, D. (2005) 'The Evolution of Morality', in D. M. Buss (ed.) *The Handbook of Evolutionary Psychology*, pp. 747–71 (Hoboken, NJ: John Wiley & Sons).

Kurzban, R. (2010) 'Grand Challenges of Evolutionary Psychology', *Frontiers in Psychology* 1:3.

Kurzban, R. (2011) *Why Everyone (Else) is a Hypocrite: Evolution and the Modular Mind* (Princeton, NJ: Princeton University Press).

Kurzban, R. and C. A. Aktipis (2007) 'On Detecting the Footprints of Multilevel Selection in Humans', in S. Gangestad and J. Simpson (eds) *The Evolution of Mind: Fundamental Questions and Controversies*, pp. 226–32 (New York: Guilford).

Kurzban, R., Tooby, J. and L. Cosmides (2001) 'Can Race be Erased?: Coalitional Computation and Social Categorization', *Proceedings of the National Academy of Sciences* 98:15387–92.

Laland, K. N. (2007) 'Review Symposium: Life After Evolutionary Psychology', *Metascience* 16:1–24.

Laland, K. N. and G. R. Brown (2002) *Sense and Nonsense: Evolutionary Perspectives on Human Behaviour*, 1st ed. (Oxford: Oxford University Press).

Laland, K. N. and G. R. Brown (2011a) *Sense and Nonsense: Evolutionary Perspectives on Human Behaviour*, 2nd ed (Oxford: Oxford University Press).

Laland, K. N. and G. R. Brown (2011b) 'The Future of Evolutionary Psychology', in V. Swami (ed.) *Evolutionary Psychology: A Critical Introduction*, pp. 343–66 (Oxford: WBPS Blackwell.)

Lehmann, L., L. Keller, S. West, D. Roze (2007), 'Group Selection and Kin Selection: Two Concepts but one Process', *Proceedings of the National Academy of Sciences of the United States of America* 104:6736–9.

Lenneberg, E. (1967) *Biological Foundations of Language* (New York: Wiley).

Lewens, T. (2007) *Darwin* (London and New York: Routledge).

Lewontin, R. C. (1990) 'The Evolution of Cognition: Questions We Will Never Answer', D. N. Osherson and E. E. Smith (eds) *An Invitation to Cognitive Science: Thinking*, pp. 229–46 (Cambridge, MA: MIT Press).

Lickliter, R. and H. Honeycutt (2003) 'Developmental Dynamics: Toward a Biologically Plausible Evolutionary Psychology', *Psychological Bulletin* 129:819–35.

Liddle, J. R. and T. K. Shackelford (2009) 'Why Evolutionary Psychology is "True"', *Evolutionary Psychology* 7:288–94.

Lieberman, D., E. G. Pillsworth and M. G. Haselton (2011) 'Kin Affiliation Across the Ovulatory Cycle: Females Avoid Fathers when Fertile', *Psychological Science* 22:13–18.

Lipton, P. (1991) *Inference to the Best Explanation* (New York: Routledge).

Lloyd, E. A. and M. W. Feldman (2002) 'Evolutionary Psychology: A View From Evolutionary Biology', *Psychological Inquiry* 13:150–6.

Machery, E. and H. C. Barrett (2006) 'Essay Review: Debunking *Adapting Minds*', *Philosophy of Science* 73:232–46.

Machery, E. and K. Cohen (2012) 'An Evidence-Based Study of the Evolutionary Behavioral Sciences', *The British Journal for the Philosophy of Science* 63:177–226.

Marshall, J. A. (2011) 'Group Selection and Kin Selection: Formally Equivalent Approaches', *Trends in Ecology and Evolution* 26:325–32.

Maynard Smith, J. (1964), 'Group Selection and Kin Selection', *Nature* 201:1145–7.

Maynard Smith, J. and G. R. Price (1973) 'The Logic of Animal Conflict', *Nature* 246:15–18.

Mekel-Bobrov, N., S. L. Gilbert, P. D. Evans, E. J. Vallender, J. R. Anderson, R. R. Hudson, *et al.* (2005) 'Ongoing Adaptive Evolution of *ASPM*, a Brain Size Determinant in *Homo sapiens*', *Science* 309:1720–2.

Meyer, M., M. Kircher, M. T. Gansauge, H. Li, F. Racimo, S. Mallick, *et al.* (2012) 'A High-Coverage Genome Sequence from an Archaic Denisovan Individual', *Science* 338;222–6.

Miller, G. (2012) 'The Smartphone Psychology Manifesto', *Perspectives on Psychological Science* 7:221–37.

Miller, S. L. and J. K. Maner (2008) 'Coping with Romantic Betrayal: Sex Differences in Responses to Partner Infidelity', *Evolutionary Psychology* 6:413–26.

Miller, G., J. Tybur and B. D. Jordan (2007) 'Ovulatory Cycle Effects on Tip Earnings by Lap Dancers', *Evolution & Human Behavior* 28:375–81.

Mithen, S. (2007) 'How the Evolution of the Human Mind Might Be Reconstructed', in S. Gangestad and J. Simpson (eds) *The Evolution of Mind: Fundamental Questions and Controversies*, pp. 60–6 (New York: Guilford).

Musgrave, A. (1974) 'Logical Versus Historical Theories of Confirmation', *British Journal for the Philosophy of Science* 25:1–23.

Nelkin, D. (2000) 'Less Selfish Than Sacred?: Genes and the Religious Impulse in Evolutionary Psychology', in H. Rose and S. Rose (eds) *Alas Poor Darwin: Arguments Against Evolutionary Psychology*, pp. 17–32 (New York: Harmony Books).

Nettle, D., D. A. Coall and T. E. Dickins (2010) 'Birthweight and Paternal Involvement Predict Early Reproduction in British Women: Evidence From the National Child Development Study', *American Journal of Human Biology* 22:172–9.

Nicholson, N. (2012) 'The Evolution of Business and Management', in S. C. Roberts (ed.) *Applied Evolutionary Psychology*, pp. 16–35 (Oxford: Oxford University Press).

O'Connor, J. J. M., D. E. Re and D. R. Feinberg (2011) 'Voice Pitch Influences Perceptions of Sexual Infidelity' *Evolutionary Psychology* 9:64–78.

Öhman, A. and S. Mineka (2001) 'Fear, Phobias and Preparedness: Toward an Evolved Module of Fear and Fear Learning', *Psychological Review* 108:483–522.

Okasha, S. (2007) 'Rational Choice, Risk Aversion, and Evolution', *The Journal of Philosophy* 104:217–35.

Okasha, S. (2008a) 'Biological Altruism', *The Stanford Encyclopedia of Philosophy*, Winter 2008 ed., available at: http://plato.stanford.edu/archives/win2008/entries/altruism-biological/ (accessed 29 October 2014).

Okasha, S. (2008b) *Evolution and the Levels of Selection* (Oxford: Oxford University Press).

Oum, R. E., D. Lieberman and A. Aylward (2011) 'A Feel for Disgust: Tactile Cues to Pathogen Presence', *Cognition and Emotion* 25:717–25.

Oyama, S. (2001) 'Terms and Tension: What do you do when all the Good Words are Taken?', in S. Oyama, P. E. Griffiths and R. D. Gray (eds) *Cycles of Contingency: Developmental Systems and Evolution*, pp. 177–93 (Cambridge, MA: MIT Press).

Oyama S., P. E. Griffiths and R. D. Gray (2001) 'Introduction: What is Developmental Systems Theory?' in S. Oyama, P. E. Griffiths and R. D. Gray (eds) *Cycles of Contingency: Developmental Systems and Evolution*, pp. 1–11 (Cambridge, MA: MIT Press).

Panksepp, J. and J. B. Panksepp (2000) 'The Seven Sins of Evolutionary Psychology', *Evolution and Cognition* 6:108–31.

Park, J. H. (2012) 'Evolutionary Perspectives on Intergroup Prejudice: Implications for Promoting Tolerance', in S. C. Roberts (ed.) *Applied Evolutionary Psychology*, 186–200 (Oxford: Oxford University Press).

Pawlowski, B., and R. I. M. Dunbar (1999a) 'Impact of Market Value on Human Mate Choice Decisions', *Proceedings of the Royal Society B: Biological Sciences* 266:281–5.

Pawlowski, B. and R. I. M. Dunbar (1999b) 'Withholding Age as Putative Deception in Mate Search Tactics', *Evolution and Human Behavior* 20:53–69.

Pawlowski, B. and S. Koziel (2002) 'The Impact of Traits Offered in Personal Advertisements', *Evolution and Human Behavior* 23:139–49.

Penton-Voak, I. S. and D. I. Perrett (2000) 'Female Preference for Male Faces Changes Cyclically: Further Evidence', *Evolution and Human Behavior* 21:39–48.

Penton-Voak, I. S., D. I. Perrett, D. L. Castles, T. Kobayashi, D. M. Burt, L. K. Murray and R. Minamisawa (1999) 'Menstrual Cycle Alters Face Preference', *Nature* 399:741–2.

Perrett, D. I., D. M. Burt, I. S. Penton-Voak, K. J. Lee, D. A. Rowland and R. Edwards (1999) 'Symmetry and Human Facial Attractiveness', *Evolution and Human Behavior* 20:295–307.

Petersen, M. B. (2012) 'The Evolutionary Psychology of Mass Politics', in S. C. Roberts (ed.) *Applied Evolutionary Psychology*, pp. 115–30 (Oxford: Oxford University Press).

Pinel, J. P. J. (2008) *Biopsychology* (Boston, MA: Pearson Education).

Pinker, S. (1994) *The Language Instinct* (New York: Harper Perennial).

Pinker, S., (1997a), 'Evolutionary Psychology: An Exchange', *The New York Review of Books* 44:56–8.

Pinker, S. (1997b) *How The Mind Works* (New York: Norton).

Pipitone, R. N. and G. G. Gallup (2008) 'Women's Voice Attractiveness Varies Across the Menstrual Cycle', *Evolution & Human Behavior* 29:268–74.

Platek, S. M. and T. K. Shackelford (eds) (2009) *Foundations in Evolutionary Cognitive Neuroscience* (Cambridge: Cambridge University Press).

Price, G. R. (1970) 'Selection and Covariance', *Nature* 227:520–1.

Profet, M. (1992) 'Pregnancy Sickness as Adaptation: A Deterrent to Maternal Ingestion of Teratogens', in J. H. Barklow, L. Cosmides and J. Tooby (eds) *The Adapted Mind: Evolutionary Psychology and the Generation of Culture*, pp. 327–65 (New York: Oxford University Press).

Puts, D. A. (2005) 'Mating Context and Menstrual Phase Affect Women's Preferences for Male Voice Pitch', *Evolution & Human Behavior* 26:388–97.

Ramachandran, V. S. (1997) 'Why do Gentlemen Prefer Blondes?', *Medical Hypotheses* 48:19–20.

Rasmussen, M., X. Guo, Y. Wang, K. E. Lohmueller, S. Rasmussen, A. Albrechtsen, *et al.* (2011), 'An Aboriginal Australian Genome Reveals Separate Human Dispersals into Asia', *Science* 334:94–8.

Richardson, R. C. (2007) *Evolutionary Psychology as Maladapted Psychology* (Cambridge, MA: MIT Press).

Roberts, S. C. and J. Havlicek (2012) 'Evolutionary Psychology and Perfume Design', in S. C. Roberts (ed.) *Applied Evolutionary Psychology*, pp. 330–48 (Oxford: Oxford University Press).

Rosen, D. E. (1982) 'Teleostean Interrelationships, Morphological Function and Evolutionary Inference', *American Zoologist* 22:261–73.

Samuels, R. (1998) 'Evolutionary Psychology and the Massive Modularity Hypothesis', *British Journal of Philosophy of Science* 49:575–602.

Sansom, R. (2003), 'Constraining the Adaptationism Debate', *Biology and Philosophy* 18: 493–512.

Schmidt, K. and J. F. Cohn (2001) 'Human Facial Expressions as Adaptations: Evolutionary Questions in Facial Expression Research', *American Journal of Physical Anthropology* 116:3–24.

Schmitt, D. P and J. J. Pilcher (2004) 'Evaluating Evidence of Psychological Adaptation: How do we know one when we see one?', *Psychological Science* 15:643–9.

Schmitt, D. P, L. Alcalay, J. Allik, L. Ault, I. Austers, K. L. Bennett, *et al.* (2003) 'Universal Sex Differences in the Desire for Sexual Variety: Tests From 52 Nations, 6 Continents, and 13 Islands', *Journal of Personality and Social Psychology* 85:85–104.

Schulz, A. (2008) 'Risky Business: Evolutionary Theory and Human Attitudes Towards Risk—A Reply to Okasha', *Journal of Philosophy* 105:156–65.

Schulz, A. (2011) 'Heuristic Evolutionary Psychology', in K. Plaisance and T. Reydon (eds) *Philosophy of Behavioral Biology*, pp. 217–34 (Berlin: Springer).

Schützwohl, A. (2008), 'Relief Over the Disconfirmation of the Prospect of Sexual and Emotional Infidelity', *Personality and Individual Differences* 44:668–78.

Sear, R., D. W. Lawson and T. E. Dickins (2007) 'Synthesis in the Human Evolutionary Behavioural Sciences', *Journal of Evolutionary Psychology* 5:3–28.

Segerstråle, U. (2001) *Defenders of the Truth: The Sociobiology Debate* (Oxford: Oxford University Press).

Sell, A., J. Tooby and L. Cosmides (2009) 'Formidability and the Logic of Human Anger', *Proceedings of the National Academy of Sciences of the United States of America* 106: 15073–8.

Shackelford, T. K. (2003) 'Preventing, Correcting and Anticipating Female Infidelity: Three Adaptive Problems of Sperm Competition', *Evolution and Cognition* 9:90–6.

Shackelford, T. K., D. M. Buss and K. Bennett (2002) 'Forgiveness or Breakup: Sex Differences in Responses to a Partner's Infidelity', *Cognition and Emotion* 16:299–307.

Shanahan, T. (2004) *The Evolution of Darwinism: Selection, Adaptation, and Progress in Evolutionary Biology* (Cambridge: Cambridge University Press).

Sharpe, K. and L. Van Gelder (2005) 'Techniques for Studying Finger Flutings', *Bulletin of Primitive Technology* 30:68–74.

Sharpe, K. and L. Van Gelder (2006) 'Evidence for Cave Marking by Paleolithic Children', *Antiquity* 80:937–47.

Shaw, C. N., C. L. Hofmann, M. D. Petraglia, J. T. Stock and J. S. Gottschall (2012) 'Neandertal Humeri may Reflect Adaptation to Scraping Tasks, but not Spear Thrusting', *PLoS ONE* 7:1–8.

Shea, N. (2012) 'New Thinking, Innateness and Inherited Representation', *Philosophical Transactions of the Royal Society B: Biological Sciences* 367:2234–44.

Simoons, F. (1969) 'Primary Adult Lactose Intolerance and the Milking Habit: A Problem in Biologic and Cultural Interrelations: I. Review of the Medical Research', *The American Journal of Digestive Diseases* 14:819–36.

Simoons, F. (1970) 'Primary Adult Lactose Intolerance and the Milking Habit: A Problem in Biologic and Cultural Interrelations: II. A Culture Historical Hypothesis', *The American Journal of Digestive Diseases* 15:695–710.

Skyrms, B. (1996) *The Evolution of the Social Contract* (Cambridge: Cambridge University Press).

Sober, E. (1993) *The Nature of Selection: Evolutionary Theory in Philosophical Focus* (Chicago, IL: University of Chicago Press).

Sober, E. (2000) *Philosophy of Biology* (Boulder, CO, and Oxford: Westview Press).

Sober, E. and D. S. Wilson (1999) *Unto Others: The Evolution and Psychology of Unselfish Behavior* (Cambridge, MA: Harvard University Press).

Somit, A. and S. A. Peterson (eds) (2003) *Human Nature and Public Policy: An Evolutionary Approach* (New York: Palgrave Macmillan).

Stearns, S. C. (1986) 'Natural Selection and Fitness, Adaptation and Constraint', in D. M. Raup and D. Jablonski (eds) *Patterns and Processes in the History of Life*, pp. 23–44 (Berlin, Heidelberg: Springer-Verlag).

Sterelny, K. (1995), 'The Adapted Mind', *Biology and Philosophy* 10:365–80.

Sterelny, K. (2007) 'An Alternative Evolutionary Psychology?', in S. Gangestad and J. Simpson (eds) *The Evolution of Mind: Fundamental Questions and Controversies*, pp. 178–85 (New York: Guilford).

Sterelny, K. and Griffiths, P.E. (1999) *Sex and Death: An Introduction to Philosophy of Biology* (Chicago, IL: University of Chicago Press).

Swami, V. (ed.) (2011) *Evolutionary Psychology: A Critical Introduction* (Oxford: WBPS Blackwell).

Swami, V. and N. Salem (2011) 'The Evolutionary Psychology of Human Beauty', in V. Swami (ed.) *Evolutionary Psychology: A Critical Introduction*, pp. 131–82 (Oxford: WBPS Blackwell).

Symons, D. (1989) 'A Critique of Darwinian Anthropology', *Ethology and Sociobiology* 10:131–44.

Symons, D. (1992) 'On the Use and Misuse of Darwinism in the Study of Human Behavior' in J. Barkow, L. Cosmides, and J. Tooby (eds) *The Adapted Mind*, pp. 137–59 (New York: Oxford University Press).

Symons, D. (2005) 'Adaptationism and Human Mating Psychology', in D. Buss (ed.) *A Handbook of Evolutionary Psychology*, pp. 255–7 (Hoboken, NJ: Wiley).

Symons, D, (2008), 'Featured Interview', *HBES Winter 2008 Newsletter*, 5–9.

Symons, D., and B. Ellis (1989) 'Human Male–Female Differences in Sexual Desire', in A. E. Rasa, C. Vogel and E. Voland (eds) *The Sociobiology of Sexual and Reproductive Strategies*, pp. 131–46 (New York: Chapman and Hall).

Sznycer, D. (2012) 'Cross-cultural Differences and Similarities in Proneness to Shame: An Adaptationist and Ecological Approach', *Evolutionary Psychology* 10:352–70.

Takahashi, H. and Y. Okubo (2009) 'Sex Differences in the Neural Correlates of Jealousy', in S. M. Platek and T. K. Shackelford (eds) *Foundations in Evolutionary Cognitive Neuroscience*, pp. 205–15 (Cambridge: Cambridge University Press).

Thornhill, R. (1997) 'The Concept of an Evolved Adaptation', in G. R. Bock and G. Cardew (eds) *Characterizing Human Psychological Adaptations*, pp. 4–13 (Chichester: Wiley).

Thornhill, R., and S. W. Gangestad (1999) 'The Scent of Symmetry: A Human Sex Pheromone That Signals Fitness?', *Evolution and Human Behavior* 20:175–201.

Thornhill, R. and N. W. Thornhill (1992) 'The Evolutionary Psychology of Men's Coercive Sexuality', *Behavioral and Brain Sciences* 15:363–421.

Thornhill, R. and C. Palmer (2000) *A Natural History of Rape: Biological Bases of Sexual Coercion* (Cambridge, MA: MIT Press).

Tinbergen, N. (1963) 'On Aims and Methods in Ethology', *Zeitschrift für Tierpsychologie* 20:410–33.

Tooby, J. and L. Cosmides (1990) 'The Past Explains the Present: Emotional Adaptations and the Structure of Ancestral Environments', *Ethology and Sociobiology* 11:375–424.

Tooby, J. and L. Cosmides (1992) 'The Psychological Foundations of Culture', in J. H. Barklow, L. Cosmides and J. Tooby (eds) *The Adapted Mind: Evolutionary Psychology and the Generation of Culture*, pp. 19–136 (New York: Oxford University Press).

Tooby J. and L. Cosmides (1995) 'Foreword', in S. Baron-Cohen (ed.) *Mindblindness: An Essay on Autism and Theory of Mind*, pp. xi–xviii (Cambridge, MA: MIT Press).

Tooby, J. and L. Cosmides (2005) 'Conceptual Foundations of Evolutionary Psychology', in D. Buss (ed.) *A Handbook of Evolutionary Psychology*, pp. 5–67 (Hoboken, NJ: Wiley).

Tooby, J. and L. Cosmides (2007) 'Evolutionary Psychology, Ecological Rationality, and the Unification of the Behavioral Sciences', *Behavioral and Brain Sciences* 30:42–3.

Tooby, J. and L. Cosmides (2008) 'The Evolutionary Psychology of the Emotions and Their Relationship to Internal Regulatory Variables', in M. Lewis (ed.) *Handbook of Emotions*, pp. 114–37 (New York: Guilford).

Tooby, J., L. Cosmides and H. C. Barrett (2005) 'Resolving the Debate on Innate Ideas: Learnability Constraints and the Evolved Interpenetration of Motivational and Conceptual Functions', in P. Carruthers, S. Laurence and

S. Stich (eds) *The Innate Mind: Structure and Content*, pp. 305–37 (New York: Oxford University Press).

Trivers, R. (1971) 'The Evolution of Reciprocal Altruism', *Quarterly Review of Biology* 46:35–57.

Trivers, R. (1985) *Social Evolution* (Menlo Park, CA: Benjamin/Cummings Pub. Co.).

Trivers, R. (2011) *Deceit and Self-Deception: Fooling Yourself the Better to Fool Others* (London: Allen Lane).

Tybur, J. M., G. F. Miller and S. W. Gangestad (2007) 'Testing the Controversy: An Empirical Examination of Adaptationists' Attitudes toward Politics and Science', *Human Nature* 18:313–28.

Wallace, B. (2010) *Getting Darwin Wrong: Why Evolutionary Psychology won't Work* (Exeter: Imprint Academic).

Wason, P. C. (1966) 'Reasoning', in B. M. Foss (ed.) *New Horizons in Psychology* (Harmondsworth: Penguin).

Waynforth, D. (1998) 'Fluctuating Asymmetry and Human Male Life-history Traits in Rural Belize', *Proceedings of the Royal Society B: Biological Sciences* 265:1497–501.

West, S. A., A. S. Griffin and A. Gardner (2007) 'Social Semantics: Altruism, Cooperation, Mutualism, Strong Reciprocity and Group Selection', *Journal of Evolutionary Biology* 20:415–32.

West, S. A., A. S. Griffin and A. Gardner (2008) 'Social Semantics: How Useful has Group Selection Been?', *Journal of Evolutionary* Biology 21:374–85.

Williams, G. C. (1966) *Adaptation and Natural Selection: A Critique of Some Current Evolutionary Thought* (Princeton, NJ: Princeton University Press).

Williams, M. A. and J. B. Mattingley (2006) 'Do Angry Men get Noticed?' *Current Biology* 6:R402–4.

Wilson, D. S. and E. O. Wilson (2007) 'Rethinking the Theoretical Foundation of Sociobiology', *The Quarterly Review of Biology* 82:327–48.

Wilson, E. O. (1975), *Sociobiology: The New Synthesis* (Cambridge, MA: Harvard University Press).

Workman, L. and W. Reader (2008) *Evolutionary Psychology: An Introduction* (Cambridge: Cambridge University Press).

Wynne-Edwards, V.C. (1962) *Animal Dispersion in Relation to Social Behaviour* (London: Oliver and Boyd).

Index